Telomeres, Diet and Human Disease

Advances and Therapeutic Opportunities

Telomeres, Diet and Human Disease

Advances and Therapeutic Opportunities

Editors

Prof. Dr. Amelia Marti del Moral

Department of Nutrition
Food Sciences and Physiology
School of Pharmacy and Nutrition
University of Navarra, Pamplona, Spain

IdiSNA, Navarra Institute for Health Research, Pamplona, Spain

Center of Biomedical Research in Physiopathology of
Obesity and Nutrition (CIBEROBN)
Institute of Health Carlos III, Madrid, Spain

and

Prof. Dr. Guillermo Zalba Goñi

Department of Biochemistry and Genetics, School of Sciences
University of Navarra, Pamplona, Spain

IdiSNA, Navarra Institute for Health Research, Pamplona, Spain

CRC Press
Taylor & Francis Group
Boca Raton London New York

CRC Press is an imprint of the
Taylor & Francis Group, an **informa** business
A SCIENCE PUBLISHERS BOOK

Cover illustration reproduced by kind courtesy of Prof. Jean Woo, Dr. Ruby Yu and Dr. Nelson Tang

CRC Press
Taylor & Francis Group
6000 Broken Sound Parkway NW, Suite 300
Boca Raton, FL 33487-2742

Version Date: 20170620

International Standard Book Number-13: 978-1-4987-5091-2 (Hardback)

Library of Congress Cataloging-in-Publication Data

Names: Marti del Moral, Amelia, editor. | Goñi, Guillermo Zalba, editor.
Title: Telomeres, diet, and human disease : advances and therapeutic opportunities / editors, Amelia Marti del Moral, Dpto. Ciencias de la Alimentacioìn y Fisiologiìa, School of Pharmacy and Nutrition, University of Navarra, Pamplona, Navarra, Spain, and Guillermo Zalba Goñi, Department of Biochemistry and Genetics, University of Navarra, Pamplona, Spain.
Description: Boca Raton, FL : CRC Press, 2017. | "A science publishers book." | Includes bibliographical references and index.
Identifiers: LCCN 2017021439| ISBN 9781498750912 (hardback : alk. paper) | ISBN 9781498750929 (e-book : alk. paper)
Subjects: LCSH: Telomere. | Chromosomes.
Classification: LCC QH600.3 .T4555 2017 | DDC 572.8/7--dc23
LC record available at https://lccn.loc.gov/2017021439

Visit the Taylor & Francis Web site at
http://www.taylorandfrancis.com

and the CRC Press Web site at
http://www.crcpress.com

Preface

This volume includes 11 contributions from an international panel of experts which cover from basic science to clinical relevant work concerning Human Telomere Homeostasis.

The major aim of the present book is to provide an in depth review of our current knowledge on human telomere integrity in regard to the risk of major diseases and also examine the influence of lifestyle choices. It also illustrates future directions in research on ageing, nutrition, telomeres, and relevant diseases such as cancer, and cardiovascular diseases.

The first part of the volume focuses on general aspects of telomere biology, mainly the molecular mechanisms underlying the telomere machinery, the role of telomerase activity and the current methods to measure telomere length.

A review of the evidence on Telomere length, aging and disease risk is reported in the second part. Since telomere length is considered as a marker of cellular and biological aging, several epidemiological studies have investigated the association between telomere length and aging-related diseases.

The third part is devoted to the influence of lifestyle choice on telomere integrity. It is well known that, environmental factors can modify telomere length and this area is receiving major attention in disease prevention and treatment. Successful lifestyle interventions could lessen telomere attrition, like those based on Mediterranean diet.

The fourth part deals with clinical aspect relevant to telomere maintenance and therapeutic opportunities. Chronic oxidative stress, together with a reduction in anti-ageing molecules (sirtuins), accelerates the ageing process and is involved in the pathophysiology of chronic diseases. Understanding these molecular mechanisms has provided evidence for novel therapeutic targets that may slow the ageing process. Dysfunction in telomere maintenance pathways plays a role in aging, cancer, atherosclerosis and other diseases. This has led to telomere maintenance as a prime target for patient therapies.

As demonstrated in this volume, the inter-disciplinary approach has yielded a rich harvest of basic knowledge concerning telomere integrity and will provide the seeds for future breakthroughs in clinical progress. The text is intended to furnish the reader with a general view of the state of the art in this novel area of research. We will be satisfied if the multidisciplinary nature of these proceedings informs and stimulates the readers. It is our hope that the volume will encourage further research and understanding of all aspects of this intriguing and complex telomere field.

The editors express their gratitude to the authors of the articles and to CRC Press for having made possible the publication of this volume.

Amelia Marti del Moral
Guillermo Zalba Goñi

Contents

Chapter 1

Homeostasis of DNA Integrity

Marcela Segatto[1] and *Maria Isabel Nogueira Cano*[2,*]

INTRODUCTION

Homeostasis comprises the tendency of an organism or cell to regulate their chemical processes that take place internally so as to maintain health, vital functions and stability, owing to the coordinated response of its parts to any situation or stimulus tending to disturb its normal condition or function. But how do cells work in order to preserve DNA homeostasis? The diverse stimuli that affect DNA integrity and stability, such as changes in the genome and in gene expression can disrupt the stable state of the cell with repercussions in pathways that regulate apoptosis, senescence, and cancer. In this chapter, we review the principal mechanisms involved in DNA homeostasis, pointing our interest in the role that telomere structure and its regulation play in this field.

Cellular Homeostasis and the Damaging Factors that Interfere with DNA Integrity and Stability

Because DNA is the repository of genetic information in each living cell, maintaining its integrity and stability is essential to life. Over their lives, organisms are exposed to many different environmental and internal stimuli that affect or modify their functionality. In fact, it has been estimated that one genome can suffer up to ten of thousands of changes per day (Lindahl and Nyberg 1972). Cells reply to DNA damage by activating complex signaling networks that decide cell destiny, supporting not only DNA repair and survival but also cell death. This decision depends on factors that are involved in DNA injure recognition, DNA repair and damage tolerance, as well as on elements involved in the activation of apoptosis, necrosis, autophagy and senescence (Roos et al. 2016) (Fig. 1).

[1] Faculdade Brasileira/Multivix Vitória, Rua José Alves 135, 29075-080, Vitoria, Espirito Santo, Brazil.
[2] Departamento de Genética, Instituto de Biociências, Universidade Estadual Paulista Júlio de Mesquita Filho UNESP-Botucatu, Rua Professor Dr Antonio Celso Wagner Zanin, s/nº, 18618-689, Botucatu, Sao Paulo, Brazil.
* Corresponding author: mariaincano@gmail.com

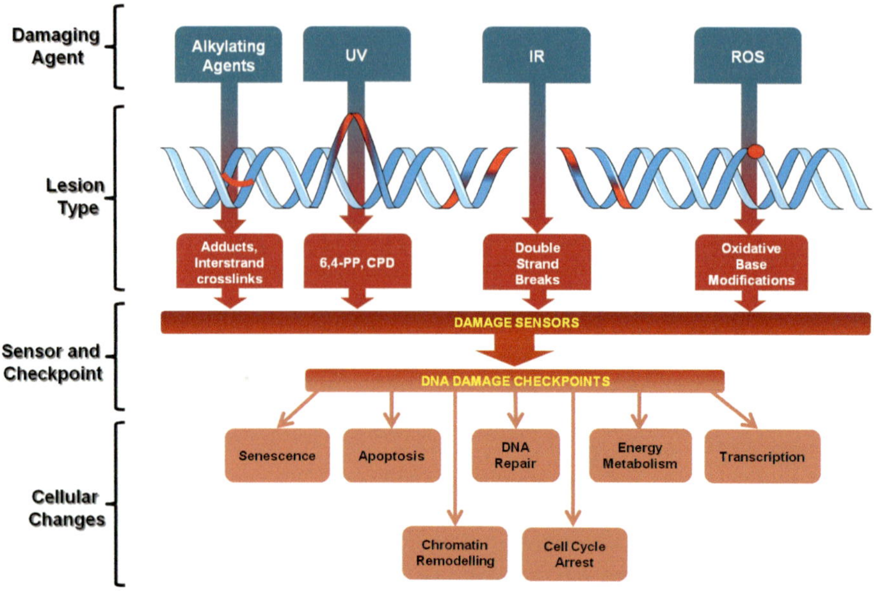

Fig. 1. DNA damage & repair: mechanisms for maintaining DNA homeostasis or induction of cell death. The genome is under constant threat since it is regularly exposed to endogenous and exogenous agents that may damage DNA. Cells have a repertoire to sense and activate DNA damage checkpoint proteins in face of different type of lesions. Ultimately, the consequences of DNA damage are activation of DNA repair pathways, of chromatin remodeling factors, modifications of transcription, fine tuning of energy metabolism, cell cycle arrest, and in case of irreparable damage, induction of senescence or apoptosis. (UV) ultraviolet radiation contained in sunlight; (IR) ionizing radiation due to, e.g., cancer therapy; (ROS) reactive oxygen species; (CPDs) cyclobutane pyrimidine dimmers; (6,4-PP) 6,4-photoproducts. (Adapted from Wolters and Schumacher 2013).

The genome is under constant threat since it is regularly exposed to exogenous agents that damage DNA. Prolonged exposure to pollutants, metals, toxic compounds, ultraviolet (UV) and ionizing radiation are some examples. Ionizing radiation typically leads to single and double-strand breaks. One example of the link between environmental-induced DNA damage and disease is that of skin cancer, caused by excessive exposure to sunlight UV radiation, which triggers two classes of DNA lesions: pyrimidine dimers and 6–4 photoproducts (Hurley 2002). These lesions distort DNA's structure impeding transcription and replication. One "hot spot" for UV-induced injure is found within the frequently mutated-tumor suppressor p53 gene, considered the "genome guardian" (Clancy 2008).

Endogenous processes can also induce DNA damage. Some examples are hydrolysis that leads to spontaneous DNA depurination, DNA alkylation can lead to adduct and interstrand crosslink formation, reactive oxygen species (ROS) induce base oxidation and DNA breaks and replication fork collapse that can result in strand breaks (Sancar et al. 2004). These DNA lesions must be repaired to prevent loss or incorrect transmission of genetic information because errors can cause abnormal development and tumorigenesis. However, which repair system to use depends on the type of lesion, type of cell and on the cell-cycle phase (Branzei 2008). The way cells do to restore DNA integrity will be discussed in the next section. Before we review how cells usually keep their genetic integrity in the replication context.

DNA Replication—General Aspects

DNA replication is a process in which all organisms must duplicate their genome with extreme accuracy before each cell division. Replication occurs in four main steps: recognition of origins of replication by different factors and assembling of the pre-replication complex in the beginning of each cell cycle will license the replication initiation at S phase, followed by the opening of the double helix and separation of the DNA strands, the priming of the template strand, and the assembly of the new DNA segment. The two strands of the double helix DNA uncoil at a specific location, known as Origin of Replication, where several enzymes and proteins will work together to prepare the strands for semi-conservative replication. Finally, special enzymes called DNA polymerases organize the assembly of the new DNA strands. This four-stage process generally applies to all cells, but specific variations may occur (Alberts 2014).

After cells divide and at the beginning of G1 phase, the pre-replication complex including the DNA helicase complex (MCM 2–7) assembles onto the DNA throughout the G1 phase (Blow and Dutta 2005). At the end of G1, cell-cycle dependent kinases trigger the initiation of DNA replication to origins of replication, which in eukaryotes are scattered throughout the genome. How cells control which origin will fire is not exactly known, because not necessarily one origin is activated every cell cycle, most of them remain dormant (McIntosh and Blow 2012). But an input from the cell cycle machinery is activated in a way that each active origin can fire only once per cell cycle and the space between them must guarantee that the whole genome is replicated during the S phase. DNA replication takes place at a Y-shaped structure named replication fork. A DNA polymerase enzyme catalyzes nucleotide polymerization in a 5'-to-3' direction, copying the parental DNA template strand with remarkable fidelity. Since the two strands of a DNA double helix are in antiparallel, this 5'-to-3' DNA synthesis can take place continuously only in one of the strands, in the leading strand, it is initiated by the addition of deoxynucleotides at the 3' end of an RNA primer synthesized by DNA primase in prokaryotes and DNA pol α/primase in eukaryotes (Chagin et al. 2010). The other strand, called the lagging strand, in contrast, is copied in short DNA fragments known as Okazaki fragments. Because the self-correcting DNA polymerase cannot start a new chain, in this case, DNA synthesis is primed by short RNA primer molecules to which deoxynucleotides are added at the 3' end. Subsequently, RNA primers are erased and replaced with DNA allowing the ligation of Okazaki fragments by many additional enzymes to create long DNA chains (Okazaki et al. 1968, Hubscher and Seo 2001).

At the molecular level, DNA replication requires the cooperation of many proteins that associate at a replication fork to form a very efficient "replication machine," through which the activities and movements of the individual components are coordinated. The replication of a genomic DNA template involves, at least: two different DNA polymerases and a DNA primase to perform nucleoside triphosphate polymerization; a DNA helicase to assist in unfolding the DNA helix and a single strand DNA-protein complex (replication protein A, RPA) to protect ssDNA during this process; RNAse H and DNA ligase respectively degrade RNA primers and seal the discontinuously synthesized lagging-strand; DNA topoisomerases, the clamp loading complex (replication factor C, RFC), and a DNA polymerase clamp or processivity factor (proliferating cell nuclear antigen, PCNA) help to relieve helical winding and DNA tangling problems. Biochemical evidence purposes the existence of large preformed multiprotein replication complexes (Noguchi et al. 1983, Tom et al. 1996). However, live microscopy evidence points to short time interactions between the components, suggesting the existence of highly dynamic complexes (Sporbert et al. 2002, Sporbert et al. 2005, Schermelleh et al. 2007, Gorisch et al. 2008).

The accuracy of copying DNA during replication is such that only about one mistake is made at every 10^9 nucleotides copied. This high fidelity, however, depends not only on complementary base-pairing but also on numerous correcting mechanisms that act sequentially if any mispairing occurs. The first proofreading step is carried out by the DNA polymerase before a new nucleotide is added to the growing chain. The enzyme must undergo a conformational change. Moreover, the right nucleotide has a higher affinity for the moving polymerase. This way allows the polymerase to "double-check" the exact base-pair geometry before it catalyzes the addition of the nucleotide (Alberts 2014).

The next error-correcting reaction takes place when an incorrect nucleotide is added. The DNA polymerase works in the 3'-to-5' direction as a proofreading exonuclease, remove any unpaired nucleotide at the primer terminus, continuing until enough residues have been removed to regenerate a base-paired with 3'-OH terminus that can prime DNA synthesis. In this way, DNA polymerase functions as a "self-correcting" enzyme that removes its own polymerization errors as it moves along the DNA (Moldovan et al. 2007). Cells have yet another chance to correct DNA polymerase self-mistakes in a process called strand-directed mismatch repair that will be discussed below.

DNA Repair Systems

DNA repair processes exist in prokaryotic and eukaryotic organisms, and many of the proteins involved are much conserved. Cells have evolved some mechanisms to detect and repair the various types of DNA damage, no matter if it is caused by the environment or by errors in cellular processes. Because DNA is a molecule that performs an active and critical role in cell division, control of DNA repair is strictly tied to regulation of the cell cycle. To guarantee that DNA is intact there are checkpoint mechanisms during the cell cycle before proceeding DNA replication and cell division. Failures in these checkpoints can increase DNA damage, which in turn leads to mutations (Clancy 2008).

All organisms have developed a complex network of DNA repair pathways: direct repair, base excision repair, nucleotide excision repair, mismatch repair, and recombination repair pathways (Fig. 2).

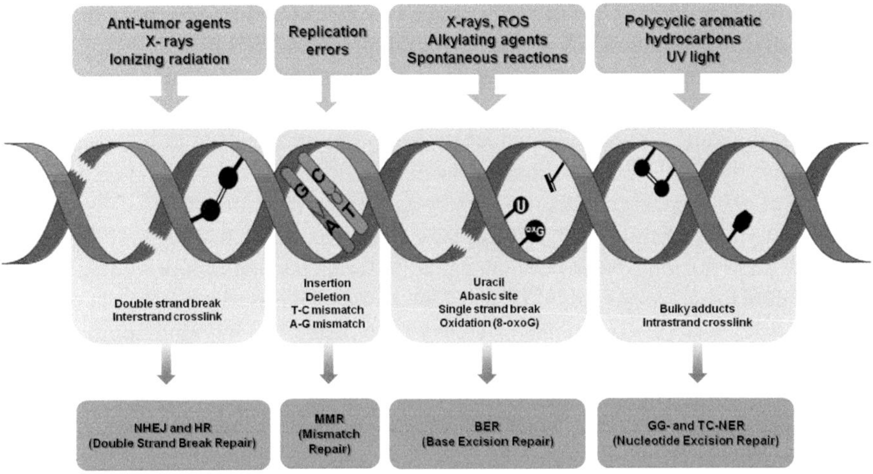

Fig. 2. DNA damaging agents, type of lesions and relevant DNA repair pathways. (ROS) reactive oxygen species; (UV) ultraviolet radiation; (8-oxoG) 8-oxoguanine; (GG-NER) global genomic nucleotide excision repair; (TC-NER) transcription-coupled nucleotide excision repair. (Adapted from Moldovan et al. 2007).

In the direct repair system, the response to DNA damage is simple to remove or reverse the lesion in one reaction, to restore the local sequence to its original state (Yu et al. 1999). From bacterial to humans several photolyases can directly reverse DNA damage resulted from UV or cisplatin treatment, in a light-dependent reaction (Fox et al. 1994, Sancar 1996, van der Spek et al. 1996, Todo et al. 1997). Another example of the direct repair enzyme is O^6-methylguanine DNA methyltransferase, which is important in the repair of alkylation adducts (Hazra et al. 2007).

Base excision repair (BER) is probably the predominant used repair pathway in the cells (Zharkov 2008, Dalhus et al. 2009). It targets small base adducts, such as those produced by methylation, oxidation, reduction, or fragmentation of bases by ionizing radiation. For example, methyl chemical groups are frequently added to guanine to form 7-methylguanine and reactive oxygen species (ROS) convert guanine to 7,8-dihydro-8-oxoguanine. Alternatively, purine or pyrimidine groups may be lost resulting in the formation of an apurinic/apyrimidinic site (AP site). All such changes generate abnormal structures thus, enzymes known as DNA glycosylases accurately recognize and remove injured bases by cutting them out of the DNA strand cleaving the covalent bonds between the bases and the sugar-phosphate backbone (this is the major feature of BER). A pair of endonucleases known as APE's incises 3' (AP lyase) and 5' (AP hydrolase) to the AP site (McCullough et al. 2001, Ischenko and Saparbaev 2002, Mundle et al. 2009, Motta et al. 2010). The resulting single-strand break can then be processed by either short-patch (where a single nucleotide is replaced) or long-patch BER (where 2–10 new nucleotides are synthesized) (Liu et al. 2007).

In the short-patch, the small nucleotide gap remained after AP site processing is filled by a specialized repair polymerase beta (pol β) and is sealed by DNA ligase III along with its cofactor XRCC1 (X-ray repair cross-complementing protein 1)

(Robertson et al. 2009). During long-patch BER, DNA strand synthesis is thought to be mediated by pol β and pol δ along with the processivity factor PCNA, the same polymerases that carry out DNA replication. Polymerase epsilon may also function in long-patch BER (Frosina et al. 1996). These polymerases perform displacing synthesis, meaning that the downstream 5' DNA end is displaced to form a flap that is refractory to ligation. Flap structure-specific endonuclease 1 (FEN1) resolves the problem of an unligatable DNA junction by catalyzing the removal of the flap. Long-patch pathway is completed by DNA ligase I (Robertson et al. 2009). The decision to proceed via the long-patch or short-patch BER mechanism is poorly understood.

While BER corrects individual or a few damaged bases, the nucleotide excision repair (NER) system corrects larger areas of DNA damage by removing the strand section that contains a significant nucleotide lesion (Moustacchi 2000, Bernstein et al. 2002, Nakano et al. 2005). NER repairs DNA lesions such as UV-induced pyrimidine dimmers and 6–4 photoproducts or more bulky adducts like cisplatin-DNA intrastrand crosslinks (Truglio et al. 2006). NER can be described as having four stages: damage recognition/preincision, incision, gap-filling, and ligation.

In prokaryotes NER is catalyzed by the product of three genes (uvrA, B, and C—for UV sensitivity). The protein UvrA identifies damaged DNA and recruits UvrB and UvrC to the lesion. UvrB and UvrC then cleave on the 3' and 5' sides of the damaged site, respectively, thus excising oligonucleotides consisting of 12 or 13 bases (Van Houten 1990, Sancar 1996). After the incision, UvrD helicase removes the nucleotide fragment, DNA polymerase I synthesizes the complementary strand, and then DNA ligase completes the repair process (Selby and Sancar 1994, Svejstrup 2002).

The molecular mechanism of NER is more complex in eukaryotes since it involves more than ten proteins (Gillet and Scharer 2006). Eukaryotic NER has two sub-pathways, global genomic repair (GGR) and transcription-coupled repair (TCR) which differ in how they recognize DNA damage. In GGR, several "damage sensing" proteins recognizes DNA lesions and begin the repair pathway, whereas, in TCR, stalling of the RNA polymerase is responsible for the initiation process (Selby and Sancar 1994, Svejstrup 2002). Then they share the same process for lesion incision, repair, and ligation.

In yeast, the proteins called RAD's (for "RADiation Sensitivity") are similar to Uvr's, such as RAD3 and RAD10 (Cooper 2000). In mammalian cells, the XPA protein (and probably also XPC) begins repair by identifying damaged DNA and forming complexes with other proteins. Among them are XPB and XPD proteins, which act as helicases that unwind the damaged DNA, and the RPA protein, which recruits the polymerases for repair synthesis allowing the complete extension and subsequent stabilization of the so-called pre-incision complex. Also, the binding of XPA to damaged DNA leads to the recruitment of XPF (as a heterodimer with ERCC1—Excision Repair Cross Complementing) and XPG to the complex. XPF/ERCC1 and XPG are endonucleases cleaving DNA on the 5' and 3' sides of the damaged site and excising about 30 bases. Then, the resultant gap appears to be filled in by DNA polymerase δ or ε (in association with RFC and PCNA) and sealed by ligase (Dexheimer 2013).

Both BER and NER systems are constantly active in the cell however, other mechanisms are activated during specific cellular stages, such as DNA mismatch repair (MMR) system which functions mainly during S phase (Jiricny 2006, Conde-Perezprina et al. 2012). The MMR machinery recognizes and corrects mismatched or unpaired bases that have escaped the proofreading activity of DNA polymerases during DNA replication. MMR proteins also correct insertion/deletion loops that result from polymerase slippage during replication of repetitive DNA sequences. The MMR pathway is divided into three principle steps: recognition of mispaired bases, excision, and gap filling (Fukui 2010).

In *E. coli*, the first steps in MMR are performed by the MutHLS system formed by proteins: MutS, MutL, and MutH (Modrich 1989, Iyer et al. 2006). In this system, a MutS homodimer recognizes and attaches to a mismatched base in the dsDNA (Takamatsu et al. 1996, Lamers et al. 2000, Obmolova et al. 2000). A MutL homodimer then interacts with and stabilizes the MutS-mismatch complex and activates a MutH restriction endonuclease (Ban and Yang 1998). The MMR system needs to discriminate the newly synthesized DNA strand to remove the incorrect base of the mismatched pair. This ability rests on the fact that *E. coli* DNA has a methylation in the adenine within the sequence GATC to form 6-methyladenine. This methylation occurs after the replication, thus newly synthesized DNA strands are not methylated and therefore can be specially recognized by the mismatch repair enzymes (Modrich 1989). At the site of a mismatch, the MutH endonuclease nicks the unmethylated strand at a hemimethylated GATC site to add an entry point for the excision reaction. The error-containing region is excised by a DNA helicase and a ssDNA-specific exonuclease. The excised tract of oligonucleotides is then replaced by DNA synthesis directed by DNA polymerase III and a ligase (Mechanic et al. 2000, Burdett et al. 2001, Yamagata et al. 2001, Yamagata et al. 2002).

Eukaryotes have a similar MMR system, although the mechanism by which eukaryotic cells identify newly replicated DNA differs from that used by prokaryotes. In mammalian cells, it seems that the strand-specificity of mismatch repair is determined by the presence of single-strand breaks (present in the newly replicated DNA) in the strand to be repaired. The eukaryotic homologs of MutS (MSH2 to MSH5) and MutL (MLH1, PMS1, and PMS2) then bind to the mismatched base and direct removal of the DNA between the strand break and the mismatch, as in *E. coli* (Modrich 2006). The degradation of the error-containing strand is performed by exonuclease 1 (Exo1) through its 3' to 5' exonucleolytic activity (Tran et al. 2004). The entry point for Exo1, which may be thousands of nucleotides from the mismatch, is generated via single-strand scission by the PCNA/RFC-dependent endonuclease activity of a MutL homolog (Kadyrov et al. 2007). The extensive gap left by Exo1 is then resynthesized by DNA polymerase δ, which is accompanied by at least two other proteins, PCNA, and RPA. Lastly, MMR is completed by DNA ligase I sealing of the remaining nick (Dexheimer 2013).

Additional forms of DNA damage are double-strand breaks (DSBs), which can be caused by ionizing radiation, nuclease dysfunction, or replication fork collapse. DNA DSBs are the most crucial lesions in DNA for inducing loss of genetic information and chromosomal instabilities. There are two different pathways to repair DSBs: homologous recombination (HR) and nonhomologous end-joining

(NHEJ) (Shrivastav et al. 2008). HR makes use of intact homologous DNA as a template for repair. The first step is initiated by a 5' to 3' degradation of DSB ends to generate 3'-ssDNA ends. Several nucleases and helicases have been involved in this step, in mammals (Huertas 2010). By contrast, most bacteria have two major sub-pathways, the RecF pathway and the RecBCD/AddAB pathway (Rocha et al. 2005, Handa et al. 2009, Yeeles and Dillingham 2010). Upon the generation of 3'-ssDNA tails, mediator proteins bind to them and load the recombinase to form a nucleoprotein filament. The recombinase searches for a homologous DNA segment and catalyzes strand invasion producing a D-loop structure. This step is catalyzed by the RecA-family recombinases (RecA in bacteria and RAD51 in eukaryotes) (Lin et al. 2006). After strand invasion, the homologous DNA is used as template for the DNA synthesis, and the intermediates are processed to form stable four-stranded DNA structures, the Holliday junctions (HJ). Capture of the second end and branch migration also occur at the same time as DNA repair synthesis. In eukaryotes, the capture of the second end appears to be mediated by RAD52 and RPA, whereas their homologs in bacteria are RecO and SSB, respectively (Kantake et al. 2002, Sugiyama et al. 2006, Nimonkar et al. 2009). Finally, HJs are resolved by endonucleases into linear duplexes, and the nicks are sealed by DNA ligase to complete the repair. Some members of a structure-specific endonuclease mammalian family are GEN1, SLX1, MUS81-EME1, and ERCC4-ERCC1, are implicated in the resolution of HJs and recombination intermediates (Svendsen and Harper 2010). In bacteria, RuvC and RusA have HJ resolvase activity (Eggleston and West 2000).

In contrast to HR, NHEJ directly ligates two DSB ends together, and although it is efficient, it is prone to loss of genetic information at the ligation sites. NHEJ is reasoned to be the predominant pathway in mammalian cells operating in all phases of the cell cycle whereas HR is restricted to the late-S and G2 phases. The initial step in the NHEJ process entails recognition and binding of the Ku heterodimer (Ku70/Ku80) to the exposed DNA termini of the DSB. Structurally, Ku adopts a preformed ring-shaped structure that completely encircles the DNA duplex (Walker et al. 2001). Ku is also involved in telomeric maintenance including regulating telomere addition and protecting telomeres from recombination and nucleolytic degradation (Boulton and Jackson 1996, Fisher and Zakian 2005). Upon binding to DNA, the Ku-DNA complex recruits the catalytic subunit of DNA-dependent protein kinases (DNA-PKcs) to generate the so-called DNA-PK holoenzyme, which exhibits protein kinase activity. The recruitment of DNA-PKcs induces an inward translocation of Ku along the DNA, allowing DNA-PKcs to contact DNA termini and to bind on opposing DSB ends promoting synapsis or tethering of the two DNA molecules. Synapsis of DNA-PKcs also results in autophosphorylation of DNA-PKcs, which allows the DNA termini to become accessible (DeFazio et al. 2002). Finally, ligation of the DNA ends is carried out by DNA ligase IV in conjunction with its binding partner XRCC4 (Critchlow et al. 1997).

The Chromosome End-Replication Problem

In 1961, Hayflick provided evidence that primary cells were not immortal, but have a restricted lifespan as they could undergo only a limited number of cell divisions

(Hayflick and Moorhead 1961). On the other side, Watson and Olovinovick named "end-replication problem" the incapacity of the DNA replication machinery to replicate the termini of linear chromosomes (telomeres), which consequently will cause gradual chromosome end shortening, at every mitotic cycle (Watson 1972, Olovnikov 1973). And Olovnikov linked the ideas that the gradual loss of chromosome ends will probably culminate in the limited capacity of cell proliferation described by Hayflick (Olovnikov 1971).

The first description of a chromosome end terminus sequence was from *Tetrahymena thermophila*, whose telomeres are composed of short tandem repeats of TTGGGG sequence ranging in size from 120–420 bp (Blackburn and Gall 1978). Ten years later, the sequence of human telomeres was described as repetitions of the TTAGGG sequence (Moyzis et al. 1988).

At that time, it was known that the semiconservative DNA replication occurs bidirectionally while DNA polymerases work only in one direction (5'-3') and requires a primer to start synthesis, which will be removed at the end of the process resulting, in about 50–200 bp of unreplicated DNA at the 3' end of all human chromosomes. The consequence of this process known as the "end-replication problem" is the progressive loss of telomeric sequences at every round of DNA replication. Other mechanisms, like processing by exo- or endonucleases and DNA damage at telomeric sites, are also responsible for progressive telomere attrition (Wu et al. 2012, Sarek et al. 2015). To counteract this continuous DNA loss, nearly all eukaryotes, with a few exceptions, use the telomerase enzyme to replenish telomeric DNA (Bonetti et al. 2014). Telomerase was discovered in the mid-1980s by Blackburn and Greider in *T. thermophila* (Greider and Blackburn 1985, Blackburn et al. 1989). This enzyme is a specialized reverse transcriptase composed by a protein component, TERT (Telomerase Reverse Transcriptase) that contains specific motifs and structurally conserved domains. TERT reverse transcribes part of the template sequence (1.5 telomeric repeat) present within its intrinsic RNA component, TER (Telomerase RNA). Telomerase thus, adds telomeric repeats to the 3'-ends of the chromosomes, elongating the G-rich telomeric strand (Greider and Blackburn 1989, Lingner and Cech 1996, Nakamura et al. 1997). In the absence of telomerase, instead, some cells can elongate telomeres by recombination-based mechanisms called Alternative Lengthening of Telomeres (ALT) (Lundblad and Blackburn 1993, Chen et al. 2001, Cesare and Reddel 2010). Chapter 2 will be completely focus on the role of telomerase in the control of telomere homeostasis.

Telomeres structure

The term "telomere" originates from the Greek: telos—end and meros—part, was introduced by Muller (Muller 1938) for the special structures, which exhibited a special heterochromatic morphology and were able to "cap" the ends of eukaryotic chromosomes. They consist of evolutionary conserved repetitive nucleotide sequences associated with specialized proteins (e.g., Shelterin in humans) (Meyne et al. 1989, Chan and Blackburn 2002). Although most telomeres share great sequence similarity, telomere length is highly variable among species, within species, in an organism, and even between chromosomes (Oeseburg et al. 2009).

Several different functions are attributed to telomeres: (i) they solve the end-replication problem of linear DNA molecules, since telomerase can elongate telomeres; (ii) they enable cells to distinguish natural chromosome ends from DNA double-strand breaks, preventing the unnecessary action of the DNA damage repair system; (iii) and they maintain chromosome stability by preventing fusion between chromosome ends (Olovnikov 1971, Watson 1972).

Telomeres from distinct organisms can form higher-order structures maintained by telomeric proteins (Griffith et al. 1999, de Lange 2004). Several lines of evidence indicate that telomeres end in a loop, known as the T-loop where the double-stranded telomeric tract curves around, allowing the single-stranded overhang to invade and pair upstream the telomeric double-stranded DNA forming a displacement loop of TTAGGG repeats called D-loop (de Lange 2004). In addition to t-loops, telomeres can form specialized structures known as G-quadruplex DNA. These structures are composed of G-quartets (or guanine tetrads), which are square planar arrays of four guanines that are hydrogen bonded by Hoogsteen base pairing (Nandakumar and Cech 2013). Two or more G-quartets can stack on top of each other to form a G-quadruplex, that may provide an alternative means of telomere protection when telomeric proteins have been displaced (Griffith et al. 1999).

In contrast to the proximal portion, the terminal region of the double-stranded telomeric DNA is not organized in nucleosomes, but is bound to specific proteins, which in humans are known as the shelterin complex (Cohen and Blackburn 1998). This complex is designed to protect the chromosomal ends from erosion and end-to-end fusion and is formed by a six-protein complex associated with telomeres (TRF1, TRF2, Rap1, TIN2, TPP1, and POT1). Telomeric repeat binding factor 1 (TRF1) and 2 (TRF2) can bind directly to double-stranded telomeric DNA (De Lange 2005). TRF1 is a negative regulator of telomere length, whereas TRF2 plays an important role in T-loop formation and telomere end protection, being considered a key player in telomere maintenance. Loss of TRF2 elicits the action of checkpoint kinases (e.g., ATM, ataxia telangectasia mutated) and the subsequent recruitment of DNA damage response components (e.g., γH2Ax and 53BP1) at chromosome ends, inducing the formation of "telomere-dysfunction induced foci" (TIF) or chromosome end-to-end fusions mediated by the non-homologous end-joining pathway (NHEJ), culminating both in genome instability. In contrast, the interaction of TRF2 with the shelterin component Rap1 (repressor activator protein 1), impairs telomere fusions by NHEJ (Feuerhahn et al. 2015).

TIN2 (TRF1-interacting nuclear factor 2) is one of the shelterin components that does not bind telomeric DNA but connects TRF1 to TRF2 and tethers TPP1 (TINF2-interacting protein 2) to TRF1 and TRF2, contributing to the stabilization of TRF2 at telomeres (Liu et al. 2004, Ye et al. 2004). Thus, TIN2 bridges the double-stranded complex (TRF1/TRF2-RAP1) to the G-rich single-stranded telomeric complex (TPP1/POT1) by its interaction with TPP1, which similarly to TIN2 does not bind telomeric DNA. TPP1, in turn, interacts with the catalytic subunit of telomerase (hTERT, human telomerase reverse transcriptase), being considered the principal telomerase recruitment factor required for telomerase activation and for telomere length regulation (Karlseder 2014, Rajavel et al. 2014). TPP1 also interacts with POT1 (protection of telomeres 1), a telomere end-binding protein that has high affinity for the G-rich single-stranded telomeric DNA. POT1 is responsible to form the D-loop in the T-loop structure and to protect telomeres; its deficiency can lead to uncapped telomeres (Wu et al. 2006).

Few years ago, homologues of the *Saccharomyces cerevisiae* CST complex were also found cohabiting telomeres of different eukaryotes, including humans. Yeast CST complex is composed by the heterotrimer Cdc13, Stn1, and Ten1, which are considered RPA-like proteins because they share structural features with the conserved heterotrimeric RPA complex. Cdc13 plays multiple roles at telomeres such as protection of the ends, telomerase activation and C-strand synthesis by its interaction with a telomerase subunit and with DNA polymerase α (Qi and Zakian 2000, Pennock et al. 2001, Grossi et al. 2004, Gao et al. 2007). In humans Cdc13 is replaced by CTC1 (conserved telomere maintenance component 1), also referred as AAF-132 (α-accessory factor) (Casteel et al. 2009), which unlike the human homologues of STN1 and TEN1, shares little sequence similarity with Cdc13 (Martin et al. 2007, Miyake et al. 2009, Surovtseva et al. 2009). But, independent of the organism, all CST complexes described so far are involved in telomere capping and C-strand filling synthesis (Casteel et al. 2009). Thus, CST complexes have been implicated in assisting DNA replication at telomeres (Miyake et al. 2009, Surovtseva et al. 2009, Gu et al. 2012). Figure 3 summarizes the coordinated interactions among the protein components of shelterin and the CST complex to control telomerase action at human telomeres.

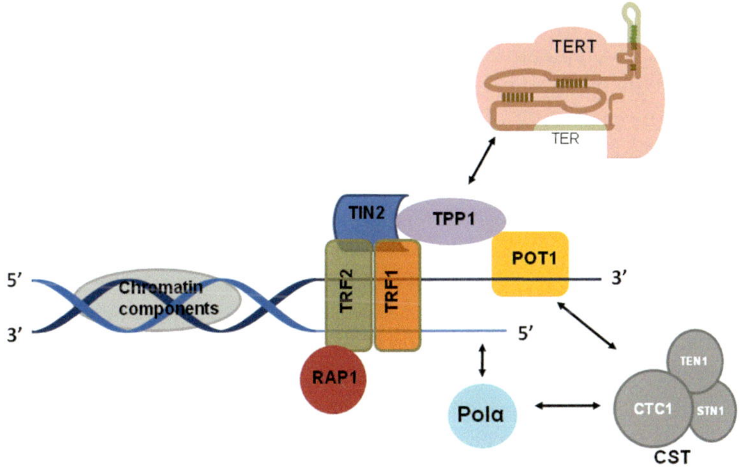

Fig. 3. Control of telomerase action and of human telomeres replication by the coordinated interaction of shelterin and the CST complex. The proteins TRF1 and TRF2 anchor the shelterin complex by interacting with the double-stranded telomeric DNA, whereas RAP1 binds directly to TRF2. The interactions between TIN2 and TPP1 bridge TRF1 and TRF2 to POT1, the G-strand telomeric DNA–binding protein. TPP1 interacts with hTERT and with POT1. POT1 interacts with the CST complex, which probably stimulates the synthesis of the telomeric C-strand DNA by DNA polymerase α-primase (Polα). Other unknown factors and general chromatin components may also help to maintain telomere homeostasis.

Disrupting Cell Homeostasis—Apoptosis, Senescence, Telomere diseases, Cancer

As previously mentioned, cells are continuously under pressure as genotoxic insults can stem from a large variety of endogenous and exogenous sources. Due to the frequency of DNA damage, to maintain homeostasis and to restore DNA integrity cells dispose of sophisticated DNA repair systems. These systems recognize specific types of lesions and induce DNA damage signaling. Failure of DNA repair has been associated with severe disorders in humans, often associated with the occurrence of cancer, premature aging, and cell death. Telomere biology is also directly involved in genome instability, biological aging, and disease processes. As telomere length shortens with age, gradual shortening of telomeres drives cells to senescence, apoptosis, oncogenic transformation of somatic cells, or increase the incidence of diseases.

Telomeres, cellular senescence and apoptosis

As described by Hayflick, primary cells in culture are not immortal; they stop dividing after a certain number of passages and become sedative, a state referred to as replicative senescence or cell aging (Hayflick and Moorhead 1961). The senescent phenotype is characterized by changes in morphology, gene expression, and proteins (Dimri et al. 1995, Shelton et al. 1999, Oeseburg et al. 2009). Multiple stimuli can cause senescence: telomere shortening, DNA damage, and induction of oncogenic or tumor suppressor signals (Chen et al. 2001, Sherr and McCormick 2002, Kurz et al. 2004). Telomere attrition is not directly involved in the acute induction of senescence, but its cumulative effect in addition to the cumulative oxidative stress might increase the probability of a cell to enter senescence (Chen et al. 2001, Kurz et al. 2004). The increasing of senescent cells in different tissues and organs may result in decreased function and pathology. On the other hand, induction of cellular senescence is an important suppressor of tumorigenesis (Shay and Wright 2002).

In a normal cellular process, telomere length decreases with age; for example, telomeres in human somatic cells shorten by 50–200 bp per cell division (Harley and Villeponteau 1995). Telomere shorter than the average size for a specific age group has been associated with increased incidence of age-related diseases and decreased lifespan in humans (Cawthon et al. 2003, Farzaneh-Far et al. 2008). The causal link between telomeres shortening and aging was proved when Bodnar et al (1998) showed that the introduction of telomerase into normal human telomerase silent cells was sufficient to bypass senescence, activate telomerase activity, and lead to cell immortalization (Bodnar et al. 1998). However, it is still under debate if we can predict life span by the absolute telomere length. For example, mice with longer telomeres do not appear to have an increased lifespan in comparison to mice with shorter telomeres (Hemann et al. 2001).

Telomere shortening has been also associated as one of the major mechanisms of replicative senescence (Vaziri et al. 1993). Thus, it is likely that telomere state and the proteins involved in the shelterin complex are critical cofactors to the induction of senescence (Karlseder et al. 2002). There is evidence that inhibition of the telomere binding proteins results in early senescence. In primary human cells, for example, expression of mutant forms of TRF2 induced growth arrest with characteristics of senescence, while overexpression of TRF2 reduces the senescence setpoint (telomere length at senescence) from 7 to 4 kb (van Steensel et al. 1998, Karlseder et al. 2002). Furthermore, inhibiting POT1 by RNA interference lead to the disappearance of telomeres 3' overhangs and induces senescence, apoptosis and chromosomal instability (Yang 2005). These results suggest that telomere capping, not just telomere length, is crucial in avoiding telomere dysfunction and preventing cell senescence disease (Oeseburg et al. 2009).

Cells that lose crucial cell cycle checkpoint functions skip this initial growth arrest (replicative senescence) and continue to divide (extended lifespan) eventually entering a second growth arrest state called crisis, when many shortened chromosome ends fuse, leading to chromosome breakage-fusion cycles almost universally driving cells to apoptosis (Shay and Wright 2010). Apoptosis, or programmed cell death, is a physiological process, involved in the regulation of tissue homeostasis in many stages of tissue development that provides a way of discarding redundant cellular material (McConkey et al. 1996, Jacobson et al. 1997, Vaux and Korsmeyer 1999). Among many structural and functional changes that occur during the induction of apoptosis, like DNA fragmentation and protein cleavage, is the modulation of telomere length and telomerase activity (Mondello and Scovassi 2004).

A rapid degradation of telomeric repeats, disruption of the telomere capping and telomerase impairment can trigger apoptotic death in several systems through different mechanisms. One mechanism relies on the inhibition of telomerase catalytic activity with the consequent telomere shortening. For example, patients affected by X-linked dyskeratosis congenita, which is characterized by altered telomerase function exhibit premature telomere shortening associated with a progressive increase of the apoptotic cellular fraction in lymphoblastoid cell lines (Montanaro et al. 2003). Mounting evidence indicates that inhibition of telomerase leads to induction of a DNA damage response and elicit an apoptotic response via p53-dependent pathway (d'Adda di Fagagna et al. 2003, Mondello and Scovassi 2004).

A second mechanism does not require telomere shortening but interferes with the telomere capping function of the telomerase protein or by chromosome uncapping through the inhibition of telomere-binding proteins. A rapid apoptosis induction was observed in different human cancer cell lines in which telomerase inhibition was achieved by targeting the RNA subunit (Saretzki et al. 2001, Cao et al. 2002, Jiang et al. 2003). This fast induction of apoptosis relies on the postulated telomere capping function of telomerase, which is independent of its telomere lengthening activity (Blackburn 2001, Sharma et al. 2003).

The third mechanism suggests a role for telomerase in the regulation of gene expression and apoptosis control (Smith et al. 2003), since in the absence of telomerase, genes involved in apoptosis could be activated (Mondello and Scovassi 2004). M14 human melanoma cells respond to the inhibition of c-Myc expression by entering crisis associated with a decrease in telomerase activity, telomere shortening, and increasing production of ROS culminating in apoptosis (Biroccio et al. 2003).

In addition, apoptotic features were visible after ~ 30 population doublings in telomerase-inhibited glioblastoma U251-MG cells (transfected with cDNA for the antisense of human telomerase RNA) when telomere shortening had probably occurred accompanied by the enhanced expression of a caspase (Kondo et al. 1998).

The faculty of inducing apoptosis in cancer cells without adversely affecting normal cell functions is desired when we think about the ideal anti-cancer drug. As one of the consequences of telomerase inhibition is to drive cells to apoptosis, nowadays, robust telomerase inhibitors are under clinical trials for the therapy of cancer (Puri 2013).

Telomeres, telomerase and cancer

As previously mentioned, human cells have two mechanisms to restrict cell growth—senescence and crisis, as both are potent anticancer protection mechanisms (Wright and Shay 1992). The normal state for most human cells is to remain in this balance between crisis period and cell death, but some rare cells acquire a mechanism, such as reactivation of telomerase expression, that can maintain or lengthen telomeres (Shay et al. 1991, Wright and Shay 1992). In this way, these rare cells can grow continuously (in other words, becomes immortal), which is believed to be a crucial step in cancer progression (Hanahan and Weinberg 2000).

Short telomere length is a risk factor for tumorigenesis (Wu et al. 2003), and this idea is supported by recent evidence (Marian et al. 2010). Progressive shortening of the telomeres will lead to activation of the DNA damage response (Hackett and Greider 2002, Wu et al. 2003). This results in the initiation of genome integrity checkpoint cascades with activation of the kinases ATM (ataxia telangiectasia mutated) and ATR (ataxia telangiectasia-and Rad3-related), and the associated downstream factors including checkpoint kinases 1 and 2 (Chk1 and Chk2), and phosphorylation of p53 (Bartek and Lukas 2003). In the case of a competent p53 pathway, senescence or apoptosis will be initiated and tumorigenesis inhibited (Deng et al. 2008). However, when the p53 pathway is defective, tumorigenesis is no longer inhibited in the presence of telomere dysfunction (Rudolph et al. 2001, Hackett and Greider 2002, Deng et al. 2008). The hallmarks of tumor development are the reactivation of telomerase and telomere stabilization; therefore, 80 to 90% of all tumors express telomerase or have a form of alternative telomere lengthening (Shay et al. 1991, Chen et al. 2000). In fact, clinical data demonstrated that telomerase overexpression in mice promoted cancer development (Gonzalez-Suarez et al. 2001, Artandi et al. 2002), while telomerase inhibition induced senescence in some cancer cells (Shammas et al. 1999).

Some clinical data revealed that telomere length (measured in lymphocytes) is shorter in patients with different types of cancer, including cancers of the head and neck, breast, bladder, prostate, lung, and kidney (Wu et al. 2003). A hypothesis to this apparent incongruence is that primary tumor cells are likely to have initially similar or shorter telomeres compared to the bulk of the tumor (Marian and Mickey 2010, Marian et al. 2010). There may thus either be an advantage and mechanism to maintain subsets of cancer cells at very short telomere lengths or the length varies as the tumor cells differentiate (Shay and Wright 2010). When telomerase is upregulated or reactivated in initiating malignant cells different outcomes are plausible: (i) may be too little telomerase is expressed, and these cells may not be able to divide long-term and they probably never become robust cancer cells; (ii) if telomerase is overexpressed, then telomeres would be predicted to grow rapidly, but long telomeres are rarely observed in primary cancers (less than 10%). Thus, there may be no selective advantage for cancer cells having more telomerase than is needed to maintain telomeres longer than that which provides protection against DNA-damage signaling and chromosome end-fusion. What is observed is that the vast majority of human cancer cells have telomeres of the same or shorter than adjacent normal tissues (Shay and Wright 2010).

Telomere associated diseases

Mutated proteins required for telomere structure, replication, repair and length maintenance have been linked to several debilitating human genetic disorders known as telomere biology disorders (TBDs). These complex diseases lead to critically short telomeres affecting the homeostasis of multiple organs. However, tissues and organs that experience frequent cell division are the most injured. The clinical manifestations and the severity of symptoms depend on the mutated gene and inherited telomere length. Besides, genome instability is commonly a hallmark of TBDs, which are associated with increased cancer risk (Sarek et al. 2015).

Dyskeratosis congenita (DC) is an inherited disease involving skin and bone marrow failure (Marrone and Mason 2003). Mutations that cause DC usually decrease telomerase complex activity (such as mutations in the dyskerin gene, TER, and TERT) or impair telomerase recruitment (as mutations in TIN2) (Heiss et al. 1998, Knight et al. 1998, Vulliamy et al. 2001, Armanios et al. 2005, Vulliamy et al. 2005, Yamaguchi et al. 2005, Savage et al. 2008, Walne et al. 2008, Yang et al. 2011). The errors normally result in defective telomere maintenance, which leads to very short telomeres, cell-cycle arrest or senescence, and stem-cell exhaustion (Montanaro et al. 2003). Patients with DC exhibit much shorter telomeres than unaffected individuals of the same age, although they are born healthy. Diagnostic features of DC include abnormal skin pigmentation, oral leukoplakia and nail dystrophy (Knight et al. 1998). Other signs that are frequently associated with DC include pulmonary fibrosis, osteoporosis, developmental delay and liver disease (Dokal 2011).

Hoyeraal-Hreidarsson (HH) syndrome is a rare and extremely severe variant of DC (Kirwan and Dokal 2009). In addition to immunodeficiency and aplastic anemia, which are the main cause of death early in their first decade, individuals with HH syndrome usually have intrauterine growth retardation, microcephaly, cerebellar hypoplasia and developmental delay (Hoyeraal et al. 1970, Hreidarsson et al. 1988, Revy et al. 2000). Similarly to the DC, the mutations associated with HH syndrome affect dyskerin, TERT, TER and TIN2, and recent reports have also linked HH dysfunction with the gene encoding RTEL1 (regulator of telomere elongation helicase 1) (Touzot et al. 2012, Ballew et al. 2013, Le Guen et al. 2013).

Aplastic anemia solely can be a manifestation of a telomere syndrome. First, DC was diagnosed as mucocutaneous features that precede aplastic anemia, now the finding of dermatologic aspects are not canonical for its diagnosis (Greider and

Blackburn 1987, Armanios et al. 2005, Armanios et al. 2007). Subsequently, mutations in hTER and hTERT genes were identified in about 3% of adults with so-called acquired aplastic anemia (Yamaguchi et al. 2003, Yamaguchi et al. 2005). Many of these patients had family members with hematologic abnormalities, suggesting that a careful trial in family history could improve aplastic anemia diagnosis in patients who carry germline mutations in the basic telomerase genes (Yamaguchi et al. 2005).

Short telomeres are also a risk factor for pulmonary fibrosis and liver cirrhosis. Idiopathic pulmonary fibrosis is the most common manifestation of TBDs. And up to 15% of pulmonary fibrosis families have germline mutations in the crucial telomerase genes (hTERT and hTER) as the causal genetic defect. Affected individuals usually develop extra-pulmonary complications related to telomere shortenings like bone marrow failure and cryptogenic liver cirrhosis (Armanios 2012).

The connection of telomerase mutations with disease in three different organ systems (blood, lungs, and liver) has important practical outcomes. In the family history, carrying even mild blood count abnormalities, pulmonary fibrosis, and hepatic cirrhosis are important clues to the diagnosis of a telomeropathy (Calado and Young 2012).

Concluding Remarks

Among the numerous of spontaneous mutations and replication errors that occur in each living cell per day, organisms sense and respond to conditions that stress their homeostasis by activating complex signaling networks that decide cell destiny. Thus, mechanisms to minimize genomic damage are essential for large long-lived species as self-correcting reactions during DNA replication, DNA injures recognition, DNA repair systems, and damage tolerance. In the absence of such mechanisms cell death and tumorigenesis would be a lot more prominent. Herein, we reviewed the factors that directly or indirectly interfere in the DNA homeostasis and we hope it has become clear the role that telomeres play in the cellular response to stress, DNA damage, cancer and disease.

In the natural course, cells have limited number of divisions as telomeres shorten with age, and progressive telomere shortening leads to senescence and apoptosis. Shorter telomeres have also been implicated in genomic instability and oncogenesis. Therefore, measuring telomere length is considered to be a good marker for human disease. Besides the length, also important to the telomere dynamics are telomerase and accessory proteins. In normal somatic cells telomerase is expressed at low levels, but it is expressed at significant levels in a wide range of cancer cells. Although, the mechanisms involved in telomerase regulation are not completely understood, novel therapeutics targeting telomerase and telomeres are of the most promising therapeutics against cancer.

Keywords: Cellular homeostasis, DNA integrity, DNA replication, DNA repair systems, telomeres, apoptosis, senescence, telomere diseases

References

Alberts, B., A. Johnson, J. Lewis, D. Morgan, M. Raff, K. Roberts et al. 2014. DNA Replication Mechanisms. Molecular Biology of the Cell. New York, Garland Science.

Armanios, M., J.L. Chen, Y.P. Chang, R.A. Brodsky, A. Hawkins, C.A. Griffin et al. 2005. Haploinsufficiency of telomerase reverse transcriptase leads to anticipation in autosomal dominant dyskeratosis congenita. Proc. Natl. Acad. Sci. USA. 102: 15960–15964.

Armanios, M. 2012. Telomerase and idiopathic pulmonary fibrosis. Mutat. Res. 730: 52–58.

Armanios, M.Y., J.J. Chen, J.D. Cogan, J.K. Alder, R.G. Ingersoll, C. Markin et al. 2007. Telomerase mutations in families with idiopathic pulmonary fibrosis. N. Engl. J. Med. 35613: 1317–1326.

Artandi, S.E., S. Alson, M.K. Tietze, N.E. Sharpless, S. Ye, R.A. Greenberg et al. 2002. Constitutive telomerase expression promotes mammary carcinomas in aging mice. Proc. Natl. Acad. Sci. USA. 99: 8191–8196.

Ballew, B.J., M. Yeager, K. Jacobs, N. Giri, J. Boland, L. Burdett et al. 2013. Germline mutations of regulator of telomere elongation helicase 1, RTEL1, in Dyskeratosis congenita. Hum. Genet. 132: 473–80.

Ban, C. and W. Yang. 1998. Structural basis for MutH activation in *E. coli* mismatch repair and relationship of MutH to restriction endonucleases. EMBO J. 17: 1526–1534.

Bartek, J. and J. Lukas. 2003. Chk1 and Chk2 kinases in checkpoint control and cancer. Cancer Cell. 3: 421–429.

Bernstein, C., H. Bernstein, C.M. Payne and H. Garewal. 2002. DNA repair/pro-apoptotic dual-role proteins in five major DNA repair pathways: fail-safe protection against carcinogenesis. Mutat. Res. 511: 145–78.

Biroccio, A., S. Amodei, A. Antonelli, B. Benassi and G. Zupi. 2003. Inhibition of c-Myc oncoprotein limits the growth of human melanoma cells by inducing cellular crisis. J. Biol. Chem. 278: 35693–35701.

Blackburn, E.H. and J.G. Gall. 1978. A tandemly repeated sequence at the termini of the extrachromosomal ribosomal RNA genes in Tetrahymena. J. Mol. Biol. 120: 33–53.

Blackburn, E.H., C.W. Greider, E. Henderson, M.S. Lee, J. Shampay and D. Shippen-Lentz. 1989. Recognition and elongation of telomeres by telomerase. Genome. 31: 553–560.

Blackburn, E.H. 2001. Switching and signaling at the telomere. Cell. 106: 661–673.

Blow, J.J. and A. Dutta. 2005. Preventing re-replication of chromosomal DNA. Nat. Rev. Mol. Cell. Biol. 6: 476–86.

Bodnar, A.G., M. Ouellette, M. Frolkis, S.E. Holt, C.P. Chiu, G.B. Morin et al. 1998. Extension of life-span by introduction of telomerase into normal human cells. Science. 279: 349–352.

Bonetti, D., M. Martina, M. Falcettoni and M.P. Longhese. 2014. Telomere-end processing: mechanisms and regulation. Chromosoma. 123: 57.

Boulton, S.J. and S.P. Jackson. 1996. Identification of a *Saccharomyces cerevisiae* Ku80 homologue: roles in DNA double strand break rejoining and in telomeric maintenance. Nucleic Acids Res. 24: 4639–4648.

Branzei, D. and M. Foiani. 2008. Regulation of DNA repair throughout the cell cycle. Nat. Rev. Mol. Cell. Biol. 9: 297–308.

Burdett, V., C. Baitinger, M. Viswanathan, S.T. Lovett and P. Modrich. 2001. *In vivo* requirement for RecJ, ExoVII, ExoI, and ExoX in methyl-directed mismatch repair. Proc. Natl. Acad. Sci. USA. 98: 6765–6770.

Calado, R. and N. Young. 2012. Telomeres in disease. Medicine Reports. 4: 8.

Cao, Y., H. Li, S. Deb and J.P. Liu. 2002. TERT regulates cell survival independent of telomerase enzymatic activity. Oncogene. 21: 3130–3138.

Casteel, D.E., S. Zhuang, Y. Zeng, F.W. Perrino, G.R. Boss, M. Goulian et al. 2009. A DNA polymerase alpha-primase cofactor with homology to replication protein A-32 regulates DNA replication in mammalian cells. J. Biol. Chem. 284: 5807–5818.

Cawthon, R.M., K.R. Smith, E. O'Brien, A. Sivatchenko and R.A. Kerber. 2003. Association between telomere length in blood and mortality in people aged 60 years or older. Lancet. 361: 393–395.

Cesare, A.J. and R.R. Reddel. 2010. Alternative lengthening of telomeres: models, mechanisms and implications. Nat. Rev. Genet. 11: 319–330.

Chagin, V.O., J.H. Stear and M.C. Cardoso. 2010. Organization of DNA replication. Cold Spring Harb Perspect Biol. 2: a000737.
Chan, S.W. and E.H. Blackburn. 2002. New ways not to make ends meet: telomerase, DNA damage proteins and heterochromatin. Oncogene. 21: 553–563.
Chen, H.J., C.L. Liang, K. Lu, J.W. Lin and C.L. Cho. 2000. Implication of telomerase activity and alternations of telomere length in the histologic characteristics of intracranial meningiomas. Cancer. 89: 2092–2098.
Chen, Q., A. Ijpma and C.W. Greider. 2001. Two survivor pathways that allow growth in the absence of telomerase are generated by distinct telomere recombination events. Mol. Cell. Biol. 21: 1819–1827.
Chen, Q.M., K.R. Prowse, V.C. Tu, S. Purdom and M.H. Linskens. 2001. Uncoupling the senescent phenotype from telomere shortening in hydrogen peroxide-treated fibroblasts. Exp. Cell. Res. 265: 294–303.
Clancy, S. 2008. DNA damage & repair: Mechanisms for maintaining DNA integrity. Nature Education. 1: 103.
Cohen, P. and E.H. Blackburn. 1998. Two types of telomeric chromatin in Tetrahymena thermophila. J. Mol. Biol. 280: 327–344.
Conde-Perezprina, J.C., M.A. Leon-Galvan and M. Konigsberg. 2012. DNA mismatch repair system: repercussions in cellular homeostasis and relationship with aging. Oxid. Med. Cell. Longev. 2012: 728430.
Cooper, G.M. and R.E. Hausman. 2000. The Cell: A Molecular Approach. DNA Repair. Sinauer Associates, Inc. Massachusetts.
Critchlow, S.E., R.P. Bowater and S.P. Jackson. 1997. Mammalian DNA double-strand break repair protein XRCC4 interacts with DNA ligase IV. Curr. Biol. 7: 588–598.
dAdda di Fagagna, F., P.M. Reaper, L. Clay-Farrace, H. Fiegler, P. Carr, T. Von Zglinicki et al. 2003. A DNA damage checkpoint response in telomere-initiated senescence. Nature. 426: 194–198.
Dalhus, B., J.K. Laerdahl, P.H. Backe and M. Bjørås. 2009. DNA base repair—recognition and initiation of catalysis. FEMS Microbiol. Rev. 33: 1044–1078.
De Lange, T. 2004. T-loops and the origin of telomeres. Nat. Rev. Mol. Cell. Biol. 5: 323–329.
De Lange, T. 2005. Telomere-related genome instability in cancer. Cold Spring Harb. Symp. Quant. Biol. 70: 197–204.
DeFazio, L.G., R.M. Stansel, J.D. Griffith and G. Chu. 2002. Synapsis of DNA ends by DNA-dependent protein kinase. EMBO J. 21: 3192–3200.
Deng, Y., S.S. Chan and S. Chang. 2008. Telomere dysfunction and tumour suppression: the senescence connection. Nat. Rev. Cancer. 8: 450–458.
Dexheimer, T.S. 2013. DNA repair pathways and mechanisms. pp. 19–32. *In*: L.A. Matheus, S.M. Cabarcas and E. Hurt [eds.]. DNA Repair of Cancer Stem Cells. Springer International publisher.
Dimri, G.P., X. Lee, G. Basile, M. Acosta, G. Scott, C. Roskelley et al. 1995. A biomarker that identifies senescent human cells in culture and in aging skin *in vivo*. Proc. Natl. Acad. Sci. USA. 92: 9363–9367.
Dokal, I. 2011. Dyskeratosis congenita. Hematology Am. Soc. Hematol. Educ. Program. 2011: 480–486.
Eggleston, A.K. and S.C. West. 2000. Cleavage of holliday junctions by the *Escherichia coli* RuvABC complex. J. Biol. Chem. 275: 26467–26476.
Farzaneh-Far, R., R.M. Cawthon, B. Na, W.S. Browner, N.B. Schiller, M.A. Whooley et al. 2008. Prognostic value of leukocyte telomere length in patients with stable coronary artery disease: data from the Heart and Soul Study. Arterioscler. Thromb. Vasc. Biol. 28: 1379–1384.
Feuerhahn, S., L.Y. Chen, B. Luke and A. Porro. 2015. No DDRama at chromosome ends: TRF2 takes centre stage. Trends Biochem. Sci. 40: 275–85.
Fisher, T.S. and V.A. Zakian. 2005. Ku: a multifunctional protein involved in telomere maintenance. DNA Repair. 4: 1215–1226.
Fox, M.E., B.J. Feldman and G. Chu. 1994. A novel role for DNA photolyase: binding to DNA damaged by drugs is associated with enhanced cytotoxicity in *Saccharomyces cerevisiae*. Mol. Cell. Biol. 14: 8071–8077.
Frosina, G., P. Fortini, O. Rossi, F. Carrozzino, G. Raspaglio, L.S. Cox et al. 1996. Two pathways for base excision repair in mammalian cells. J. Biol. Chem. 271: 9573–9578.
Fukui, K. 2010. DNA mismatch repair in eukaryotes and bacteria. J. Nucleic Acids. Article ID 260512, 1–16.

Gao, H., R.B. Cervantes, E.K. Mandell, J.H. Otero and V. Lundblad. 2007. RPA-like proteins mediate yeast telomere function. Nat. Struct. Mol. Biol. 14: 208–214.

Gillet, L.C. and O.D. Scharer. 2006. Molecular mechanisms of mammalian global genome nucleotide excision repair. Chem. Rev. 106: 253–276.

Gonzalez-Suarez, E., E. Samper, A. Ramírez, J.M. FloreS, J. Martín-Caballero, J.L. Jorcano et al. 2001. Increased epidermal tumors and increased skin wound healing in transgenic mice overexpressing the catalytic subunit of telomerase, mTERT, in basal keratinocytes. EMBO J. 20: 2619–2630.

Gorisch, S.M., A. Sporbert, J.H. Stear, I. Grunewald, D. Nowak, E. Warbrick et al. 2008. Uncoupling the replication machinery: replication fork progression in the absence of processive DNA synthesis. Cell. Cycle. 7: 1983–1990.

Greider, C.W. and E.H. Blackburn. 1985. Identification of a specific telomere terminal transferase activity in Tetrahymena extracts. Cell. 43: 405–413.

Greider, C.W. and E.H. Blackburn. 1987. The telomere terminal transferase of Tetrahymena is a ribonucleoprotein enzyme with two kinds of primer specificity. Cell. 51: 887–898.

Greider, C.W. and E.H. Blackburn. 1989. A telomeric sequence in the RNA of Tetrahymena telomerase required for telomere repeat synthesis. Nature. 337: 331–337.

Griffith, J.D., L. Comeau, S. Rosenfield, R.M. Stansel, A. Bianchi, H. Moss et al. 1999. Mammalian telomeres end in a large duplex loop. Cell. 97: 503–514.

Grossi, S., A. Puglisi, P.V. Dmitriev, M. Lopes and D. Shore. 2004. Pol12, the B subunit of DNA polymerase alpha, functions in both telomere capping and length regulation. Genes Dev. 18: 992–1006.

Gu, P., J.N. Min, Y. Wang, C. Huang, T. Peng, W. Chai et al. 2012. CTC1 deletion results in defective telomere replication, leading to catastrophic telomere loss and stem cell exhaustion. EMBO J. 31: 2309–2321.

Hackett, J.A. and C.W. Greider. 2002. Balancing instability: dual roles for telomerase and telomere dysfunction in tumorigenesis. Oncogene. 21: 619–626.

Hanahan, D. and R.A. Weinberg. 2000. The hallmarks of cancer. Cell. 100: 57–70.

Handa, N., K. Morimatsu, S.T. Lovett and S.C. Kowalczykowski. 2009. Reconstitution of initial steps of dsDNA break repair by the RecF pathway of *E. coli*. Genes Dev. 23: 1234–1245.

Harley, C.B. and B. Villeponteau. 1995. Telomeres and telomerase in aging and cancer. Curr. Opin. Genet. Dev. 5: 249–255.

Hayflick, L. and P.S. Moorhead. 1961. The serial cultivation of human diploid cell strains. Exp. Cell. Res. 25: 585–621.

Hazra, T.K., A. Das, S. Das, S. Choudhury, Y.W. Kow and R. Roy. 2007. Oxidative DNA damage repair in mammalian cells: a new perspective. DNA Repair (Amst.). 6: 470–480.

Heiss, N.S., S.W. Knight, T.J. Vulliamy, S.M. Klauck, S. Wiemann, P.J. Mason et al. 1998. X-linked dyskeratosis congenita is caused by mutations in a highly conserved gene with putative nucleolar functions. Nat. Genet. 19: 32–38.

Hemann, M.T., M.A. Strong, L.Y. Hao and C.W. Greider. 2001. The shortest telomere, not average telomere length, is critical for cell viability and chromosome stability. Cell. 107: 67–77.

Hoyeraal, H.M., J. Lamvik and P.J. Moe. 1970. Congenital hypoplastic thrombocytopenia and cerebral malformations in two brothers. Acta Paediatr. Scand. 59: 185–191.

Hreidarsson, S., K. Kristjansson, G. Johannesson and J.H. Johannsson. 1988. A syndrome of progressive pancytopenia with microcephaly, cerebellar hypoplasia and growth failure. Acta Paediatr. Scand. 77: 773–775.

Hubscher, U. and Y.S. Seo. 2001. Replication of the lagging strand: a concert of at least 23 polypeptides. Mol. Cells. 12: 149–157.

Huertas, P. 2010. DNA resection in eukaryotes: deciding how to fix the break. Nat. Struct. Mol. Biol. 17: 11–16.

Hurley, L.H. 2002. DNA and its associated processes as targets for cancer therapy. Nat. Rev. Cancer. 2: 188–200.

Ischenko, A.A. and M.K. Saparbaev. 2002. Alternative nucleotide incision repair pathway for oxidative DNA damage. Nature. 415: 183–187.

Iyer, R.R., A. Pluciennik, V. Burdett and P.L. Modrich. 2006. DNA mismatch repair: functions and mechanisms. Chem. Rev. 106: 302–323.

Jacobson, M.D., M. Weil and M.C. Raff. 1997. Programmed cell death in animal development. Cell. 88: 347–354.

Jiang, Y.A., H.S. Luo, Y.Y. Zhang, L.F. Fan, C.Q. Jiang and W.J. Chen. 2003. Telomerase activity and cell apoptosis in colon cancer cell by human telomerase reverse transcriptase gene antisense oligodeoxynucleotide. World J. Gastroenterol. 9: 1981–1984.

Jiricny, J. 2006. The multifaceted mismatch-repair system. Nat. Rev. Mol. Cell. Biol. 7: 335–346.

Kadyrov, F.A., S.F. Holmes, M.E. Arana, O.A. Lukianova, M. O'Donnell, T. Kunkel et al. 2007. *Saccharomyces cerevisiae* MutLalpha is a mismatch repair endonuclease. J. Biol. Chem. 282: 37181–37190.

Kantake, N., M.V. Madiraju, T. Sugiyama and S.C. Kowalczykowski. 2002. *Escherichia coli* RecO protein anneals ssDNA complexed with its cognate ssDNA-binding protein: A common step in genetic recombination. Proc. Natl. Acad. Sci. USA. 99: 15327–15332.

Karlseder, J., A. Smogorzewska and T. de Lange. 2002. Senescence induced by altered telomere state, not telomere loss. Science. 295: 2446–2449.

Karlseder, J. 2014. Modern genome editing meets telomeres: the many functions of TPP1. Genes Dev. 28: 1857–1858.

Kirwan, M. and I. Dokal. 2009. Dyskeratosis congenita, stem cells and telomeres. Biochim. Biophys. Acta. 1792: 371–379.

Knight, S., T. Vulliamy, A. Copplestone, E. Gluckman, P. Mason and I. Dokal. 1998. Dyskeratosis Congenita (DC) Registry: identification of new features of DC. Br. J. Haematol. 103: 990–996.

Kondo, S., Y. Tanaka, Y. Kondo, M. Hitomi, G.H. Barnett, Y. Ishizaka et al. 1998. Antisense telomerase treatment: induction of two distinct pathways, apoptosis and differentiation. FASEB J. 12: 801–811.

Kurz, D.J., S. Decary, Y. Hong, E. Trivier, A. Akhmedov and J.D. Erusalimsky. 2004. Chronic oxidative stress compromises telomere integrity and accelerates the onset of senescence in human endothelial cells. J. Cell. Sci. 117: 2417–26.

Lamers, M.H., A. Perrakis, J.H. Enzlin, H.H. Winterwerp, N. de Wind and T.K. Sixma. 2000. The crystal structure of DNA mismatch repair protein MutS binding to a G x T mismatch. Nature. 407: 711–717.

Le Guen, T., L. Jullien, F. Touzot, M. Schertzer, L. Gaillard, M. Perderiset et al. 2013. Human RTEL1 deficiency causes Hoyeraal-Hreidarsson syndrome with short telomeres and genome instability. Hum. Mol. Genet. 22: 3239–3249.

Lin, Z., H. Kong, M. Nei and H. Ma. 2006. Origins and evolution of the recA/RAD51 gene family: evidence for ancient gene duplication and endosymbiotic gene transfer. Proc. Natl. Acad. Sci. USA. 103: 10328–10333.

Lindahl, T. and B. Nyberg. 1972. Rate of depurination of native deoxyribonucleic acid. Biochemistry. 11: 3610–3618.

Lingner, J. and T.R. Cech. 1996. Purification of telomerase from Euplotes aediculatus: requirement of a primer 3 overhang. Proc. Natl. Acad. Sci. USA. 93: 10712–10717.

Liu, D., M.S. OConnor, J. Qin and Z. Songyang. 2004. Telosome, a mammalian telomere-associated complex formed by multiple telomeric proteins. J. Biol. Chem. 279: 51338–51342.

Liu, Y., R. Prasad, W.A. Beard, P.S. Kedar, E.W. Hou, D.D. Shock et al. 2007. Coordination of steps in single-nucleotide base excision repair mediated by apurinic/apyrimidinic endonuclease 1 and DNA polymerase beta. J. Biol. Chem. 282: 13532–13541.

Lundblad, V. and E.H. Blackburn. 1993. An alternative pathway for yeast telomere maintenance rescues est1-senescence. Cell. 7: 347–60.

Marian, C.O., S.K. Cho, K.J. Hatanpaa, W.E. Wright, E. Maher, C. Madden et al. 2010. The telomerase inhibitor GRN163L blocks telomerase activity. Induces telomere attrition and limits GBM stem cell proliferation and *in-vivo* growth. Clin. Cancer Res. 16: 154–63.

Marian, C.O., W.E. Wright and J.W. Shay. 2010. The effects of telomerase inhibition on prostate tumor-initiating cells. Int. J. Cancer. 127: 321–331.

Marrone, A. and P.J. Mason. 2003. Dyskeratosis congenita. Cell. Mol. Life Sci. 60: 507–517.

Martin, V., L.L. Du, S. Rozenzhak and P. Russell. 2007. Protection of telomeres by a conserved Stn1-Ten1 complex. Proc. Natl. Acad. Sci. USA. 104: 14038–14043.

McConkey, D.J., B. Zhivotovsky and S. Orrenius. 1996. Apoptosis—molecular mechanisms and biomedical implications. Mol. Aspects Med. 17: 1–110.

McCullough, A.K., A. Sanchez, M.L. Dodson, P. Marapaka, J.S. Taylor and R.S. Lloyd. 2001. The reaction mechanism of DNA glycosylase/AP lyases at abasic sites. Biochemistry. 40: 561–568.

McIntosh, D. and J.J. Blow. 2012. Dormant origins, the licensing checkpoint, and the response to replicative stresses. Cold Spring Harb. Perspect. Biol. 4(10).

Mechanic, L.E., B.A. Frankel and S.W. Matson. 2000. *Escherichia coli* MutL loads DNA helicase II onto DNA. J. Biol. Chem. 275: 38337–38346.

Meyne, J., R.L. Ratliff and R.K. Moyzis. 1989. Conservation of the human telomere sequence (TTAGGG) n among vertebrates. Proc. Natl. Acad. Sci. USA. 86: 7049–7053.

Miyake, Y., M. Nakamura, A. Nabetani, S. Shimamura, M. Tamura, S. Yonehara et al. 2009. RPA-like mammalian Ctc1-Stn1-Ten1 complex binds to single-stranded DNA and protects telomeres independently of the Pot1 pathway. Mol. Cell. 36: 193–206.

Modrich, P. 1989. Methyl-directed DNA mismatch correction. J. Biol. Chem. 264: 6597–3600.

Modrich, P. 2006. Mechanisms in eukaryotic mismatch repair. J. Biol. Chem. 281: 30305–30309.

Moldovan, G.L., B. Pfander and S. Jentsch. 2007. PCNA, the maestro of the replication fork. Cell. 129: 665–679.

Mondello, C. and A.I. Scovassi. 2004. Telomeres, telomerase, and apoptosis. Biochem. Cell. Biol. 82: 498–507.

Montanaro, L., P.L. Tazzari and M. Derenzini. 2003. Enhanced telomere shortening in transformed lymphoblasts from patients with X linked dyskeratosis. J. Clin. Pathol. 56: 583–586.

Motta, E.S., P.T. Souza-Santos, T.R. Cassiano, F.J. Dantas, A. Caldeira-de-Araujo and J.C. De Mattos. 2010. Endonuclease IV is the main base excision repair enzyme involved in DNA damage induced by UVA radiation and stannous chloride. J. Biomed. Biotechnol. 2010: 376218.

Moustacchi, E. 2000. DNA damage and repair: consequences on dose-responses. Mutat. Res. 464: 35–40.

Moyzis, R.K., J.M. Buckingham, L.S. Cram, M. Dani, L.L. Deaven, M.D. Jones et al. 1988. A highly conserved repetitive DNA sequence, (TTAGGG)n, present at the telomeres of human chromosomes. Proc. Natl. Acad. Sci. USA. 85: 6622–6626.

Muller, H.J. 1938. The remaking of chromosomes. Collecting Net. 13: 15.

Mundle, S.T., J.C. Delaney, J.M. Essigmann and P.R. Strauss. 2009. Enzymatic mechanism of human apurinic/apyrimidinic endonuclease against a THF AP site model substrate. Biochemistry. 48: 19–26.

Nakamura, T.M., G.B. Morin, K.B. Chapman, S.L. Weinrich, W.H. Andrews, J. Lingner et al. 1997. Telomerase catalytic subunit homologs from fission yeast and human. Science. 277: 955–959.

Nakano, T., A. Katafuchi, H. Terato, T. Suzuki, B. Van Houten and H. Ide. 2005. Activity of nucleotide excision repair enzymes for oxanine cross-link lesions. Nucleic Acids Symp. Ser. (Oxf.). 49: 293–294.

Nandakumar, J. and T.R. Cech. 2013. Finding the end: recruitment of telomerase to telomeres. Nat. Rev. Mol. Cell. Biol. 14: 69–82.

Nimonkar, A.V., R.A. Sica and S.C. Kowalczykowski. 2009. Rad52 promotes second-end DNA capture in double-stranded break repair to form complement-stabilized joint molecules. Proc. Natl. Acad. Sci. USA. 106: 3077–3082.

Noguchi, H., G. Prem veer Reddy and A.B. Pardee. 1983. Rapid incorporation of label from ribonucleoside disphosphates into DNA by a cell-free high molecular weight fraction from animal cell nuclei. Cell. 32: 443–451.

Obmolova, G., C. Ban, P. Hsieh and W. Yang. 2000. Crystal structures of mismatch repair protein MutS and its complex with a substrate DNA. Nature. 407: 703–710.

Oeseburg, H., R.A. de Boer, W.H. van Gilst and P. van der Harst. 2009. Telomere biology in healthy aging and disease. Pflugers Arch. 459: 259–268.

Okazaki, R., T. Okazaki, K. Sakabe, K. Sugimoto and A. Sugino. 1968. Mechanism of DNA chain growth. I. Possible discontinuity and unusual secondary structure of newly synthesized chains. Proc. Natl. Acad. Sci. USA. 59: 598–605.

Olovnikov, A.M. 1971. Principle of marginotomy in template synthesis of polynucleotides. Dokl Akad Nauk SSSR. 201: 1496–1499.

Olovnikov, A.M. 1973. A theory of marginotomy. The incomplete copying of template margin in enzymic synthesis of polynucleotides and biological significance of the phenomenon. J. Theor. Biol. 41: 181–90.

Pennock, E., K. Buckley and V. Lundblad. 2001. Cdc13 delivers separate complexes to the telomere for end protection and replication. Cell. 104: 387–96.

Puri, N. and J. Girard. 2013. Novel therapeutics targeting telomerase and telomeres. J. Cancer Sci. Ther. 5: 1–3.

Qi, H. and V.A. Zakian. 2000. The Saccharomyces telomere-binding protein Cdc13p interacts with both the catalytic subunit of DNA polymerase alpha and the telomerase-associated est1 protein. Genes Dev. 14: 1777–1788.

Rajavel, M., M.R. Mullins and D.J. Taylor. 2014. Multiple facets of TPP1 in telomere maintenance. Biochim. Biophys. Acta. 1844: 1550–1559.
Revy, P., M. Busslinger, K. Tashiro, F. Arenzana, P. Pillet, A. Fischer et al. 2000. A syndrome involving intrauterine growth retardation, microcephaly, cerebellar hypoplasia, B lymphocyte deficiency, and progressive pancytopenia. Pediatrics. 105(3): E39.
Robertson, A.B., A. Klungland, T. Rognes and I. Leiros. 2009. DNA repair in mammalian cells: Base excision repair: the long and short of it. Cell. Mol. Life Sci. 66: 981–993.
Rocha, E.P., E. Cornet and B. Michel. 2005. Comparative and evolutionary analysis of the bacterial homologous recombination systems. PLoS Genet. 1(2): e15.
Roos, W.P., A.D. Thomas and B. Kaina. 2016. DNA damage and the balance between survival and death in cancer biology. Nat. Rev. Cancer. 16: 20–33.
Rudolph, K.L., M. Millard, M.W. Bosenberg and R.A. DePinho. 2001. Telomere dysfunction and evolution of intestinal carcinoma in mice and humans. Nat. Genet. 28: 155–159.
Sancar, A. 1996. DNA excision repair. Annu. Rev. Biochem. 65: 43–81.
Sancar, A., L.A. Lindsey-Boltz, K. Unsal-Kaçmaz and S. Linn. 2004. Molecular mechanisms of mammalian DNA repair and the DNA damage checkpoints. Annu. Rev. Biochem. 73: 39–85.
Sarek, G., P. Marzec, P. Margalef and S.J. Boulton. 2015. Molecular basis of telomere dysfunction in human genetic diseases. Nat. Struct. Mol. Biol. 22: 867–874.
Saretzki, G., A. Ludwig, T. von Zglinicki and I.B. Runnebaum. 2001. Ribozyme-mediated telomerase inhibition induces immediate cell loss but not telomere shortening in ovarian cancer cells. Cancer Gene Ther. 8: 827–834.
Savage, S.A., N. Giri, G.M. Baerlocher, N. Orr, P.M. Lansdorp and B.P. Alter. 2008. TINF2, a component of the shelterin telomere protection complex, is mutated in dyskeratosis congenita. Am. J. Hum. Genet. 82: 501–509.
Schermelleh, L., A. Haemmer, F. Spada, N. Rösing, D. Meilinger, U. Rothbauer et al. 2007. Dynamics of Dnmt1 interaction with the replication machinery and its role in postreplicative maintenance of DNA methylation. Nucleic Acids Res. 35: 4301–4312.
Selby, C.P. and A. Sancar. 1994. Mechanisms of transcription-repair coupling and mutation frequency decline. Microbiol. Rev. 58: 317–329.
Shammas, M.A., C.G. Simmons, D.R. Corey and R.J. Shmookler Reis. 1999. Telomerase inhibition by peptide nucleic acids reverses immortality of transformed human cells. Oncogene. 18: 6191–200.
Sharma, G.G., A. Gupta, H. Wang, H. Scherthan, S. Dhar, V. Gandhi et al. 2003. hTERT associates with human telomeres and enhances genomic stability and DNA repair. Oncogene. 22: 131–146.
Shay, J.W., W.E. Wright and H. Werbin. 1991. Defining the molecular mechanisms of human cell immortalization. Biochim. Biophys. Acta. 1072: 1–7.
Shay, J.W. and W.E. Wright. 2002. Telomerase: a target for cancer therapeutics. Cancer Cell. 2: 257–265.
Shay, J.W. and W.E. Wright. 2010. Telomeres and telomerase in normal and cancer stem cells. FEBS Lett. 584: 3819–3825.
Shelton, D.N., E. Chang, P.S. Whittier, D. Choi and W.D. Funk. 1999. Microarray analysis of replicative senescence. Curr. Biol. 9: 939–945.
Sherr, C.J. and F. McCormick. 2002. The RB and p53 pathways in cancer. Cancer Cell. 2: 103–112.
Shrivastav, M., L.P. De Haro and J.A. Nickoloff. 2008. Regulation of DNA double-strand break repair pathway choice. Cell. Res. 18: 134–147.
Smith, L.L., H.A. Coller and J.M. Roberts. 2003. Telomerase modulates expression of growth-controlling genes and enhances cell proliferation. Nat. Cell. Biol. 5: 474–479.
Sporbert, A., A. Gahl, R. Ankerhold, H. Leonhardt and M.C. Cardoso. 2002. DNA polymerase clamp shows little turnover at established replication sites but sequential *de novo* assembly at adjacent origin clusters. Mol. Cell. 10: 1355–1365.
Sporbert, A., P. Domaing, H. Leonhardt and M.C. Cardoso. 2005. PCNA acts as a stationary loading platform for transiently interacting Okazaki fragment maturation proteins. Nucleic Acids Res. 33: 3521–3528.
Sugiyama, T., N. Kantake, Y. Wu and S.C. Kowalczykowski. 2006. Rad52-mediated DNA annealing after Rad51-mediated DNA strand exchange promotes second ssDNA capture. EMBO J. 25: 5539–5548.
Surovtseva, Y.V., D. Churikov, K.A. Boltz, X. Song, J.C. Lamb, R. Warrington et al. 2009. Conserved telomere maintenance component 1 interacts with STN1 and maintains chromosome ends in higher eukaryotes. Mol. Cell. 36: 207–18.

Svejstrup, J.Q. 2002. Mechanisms of transcription-coupled DNA repair. Nat. Rev. Mol. Cell. Biol. 3: 21–9.

Svendsen, J.M. and J.W. Harper. 2010. GEN1/Yen1 and the SLX4 complex: Solutions to the problem of Holliday junction resolution. Genes Dev. 24: 521–536.

Takamatsu, S., R. Kato and S. Kuramitsu. 1996. Mismatch DNA recognition protein from an extremely thermophilic bacterium, *Thermus thermophilus* HB8. Nucleic Acids Res. 24: 640–647.

Todo, T., H. Tsuji, E. Otoshi, K. Hitomi, S.T. Kim and M. Ikenaga. 1997. Characterization of a human homolog of (6–4) photolyase. Mutat. Res. 384: 195–204.

Tom, T.D., L.H. Malkas and R.J. Hickey. 1996. Identification of multiprotein complexes containing DNA replication factors by native immunoblotting of HeLa cell protein preparations with T-antigen-dependent SV40 DNA replication activity. J. Cell. Biochem. 63: 259–267.

Touzot, F., L. Gaillard, N. Vasquez, T. Le Guen, Y. Bertrand, J. Bourhis et al. 2012. Heterogeneous telomere defects in patients with severe forms of dyskeratosis congenita. J. Allergy Clin. Immunol. 129: 473–482.

Tran, P.T., N. Erdeniz, L.S. Symington and R.M. Liskay. 2004. EXO1-A multi-tasking eukaryotic nuclease. DNA Repair (Amst.). 3: 1549–1559.

Truglio, J.J., D.L. Croteau, B. Van Houten and C. Kisker. 2006. Prokaryotic nucleotide excision repair: the UvrABC system. Chem. Rev. 106: 233–252.

van der Spek, P.J., K. Kobayashi, D. Bootsma, M. Takao, A.P. Eker and A. Yasui. 1996. Cloning, tissue expression, and mapping of a human photolyase homolog with similarity to plant blue-light receptors. Genomics. 37: 177–182.

Van Houten, B. 1990. Nucleotide excision repair in *Escherichia coli*. Microbiol. Rev. 54: 18–51.

van Steensel, B., A. Smogorzewska and T. de Lange. 1998. TRF2 protects human telomeres from end-to-end fusions. Cell. 92: 401–413.

Vaux, D.L. and S.J. Korsmeyer. 1999. Cell death in development. Cell. 96: 245–254.

Vaziri, H., F. Schachter, I. Uchida, L. Wei, X. Zhu, R. Effros et al. 1993. Loss of telomeric DNA during aging of normal and trisomy 21 human lymphocytes. Am. J. Hum. Genet. 52: 661–667.

Vulliamy, T., A. Marrone, F. Goldman, A. Dearlove, M. Bessler, P.J. Mason et al. 2001. The RNA component of telomerase is mutated in autosomal dominant dyskeratosis congenita. Nature. 413: 432–435.

Vulliamy, T.J., A. Walne, A. Baskaradas, P.J. Mason, A. Marrone and I. Dokal. 2005. Mutations in the reverse transcriptase component of telomerase (TERT) in patients with bone marrow failure. Blood Cells Mol. Dis. 34: 257–263.

Walker, J.R., R.A. Corpina and J. Goldberg. 2001. Structure of the Ku heterodimer bound to DNA and its implications for double-strand break repair. Nature. 412: 607–614.

Walne, A.J., T. Vulliamy, R. Beswick, M. Kirwan and I. Dokal. 2008. TINF2 mutations result in very short telomeres: analysis of a large cohort of patients with dyskeratosis congenita and related bone marrow failure syndromes. Blood. 112: 3594–3600.

Watson, J.D. 1972. Origin of concatemeric T7 DNA. Nat. New Biol. 239: 197–201.

Wolters, S. and B. Schumacher. 2013. Genome maintenance and transcription integrity in aging and disease. Front Genet. 4: 19.

Wright, W.E. and J.W. Shay. 1992. The two-stage mechanism controlling cellular senescence and immortalization. Exp. Gerontol. 27: 383–389.

Wu, L., A.S. Multani, H. He, W. Cosme-Blanco, Y. Deng, J.M. Deng et al. 2006. Pot1 deficiency initiates DNA damage checkpoint activation and aberrant homologous recombination at telomeres. Cell. 126: 49–62.

Wu, P., H. Takai and T. de Lange. 2012. Telomeric 3 overhangs derive from resection by Exo1 and Apollo and fill-in by POT1b-associated CST. Cell. 150: 39–52.

Wu, X., C.I. Amos, Y. Zhu, H. Zhao, B.H. Grossman, J.W. Shay et al. 2003. Telomere dysfunction: a potential cancer predisposition factor. J. Natl. Cancer Inst. 95: 1211–1218.

Yamagata, A., R. Masui, Y. Kakuta, S. Kuramitsu and K. Fukuyama. 2001. Overexpression, purification and characterization of RecJ protein from Thermus thermophilus HB8 and its core domain. Nucleic Acids Res. 29: 4617–4624.

Yamagata, A., Y. Kakuta, R. Masui and K. Fukuyama. 2002. The crystal structure of exonuclease RecJ bound to Mn2+ ion suggests how its characteristic motifs are involved in exonuclease activity. Proc. Natl. Acad. Sci. USA. 99: 5908–5912.

Yamaguchi, H., G.M. Baerlocher, P.M. Lansdorp, S.J. Chanock, O. Nunez, E. Sloand et al. 2003. Mutations of the human telomerase RNA gene (TERC) in aplastic anemia and myelodysplastic syndrome. Blood. 102: 916–918.

Yamaguchi, H., R.T. Calado, H. Ly, S. Kajigaya, G.M. Baerlocher, S.J. Chanock et al. 2005. Mutations in TERT, the gene for telomerase reverse transcriptase, in aplastic anemia. N. Engl. J. Med. 352: 1413–1424.

Yang, D., Q. He, H. Kim, W. Ma and Z. Songyang. 2011. TIN2 protein dyskeratosis congenita missense mutants are defective in association with telomerase. J. Biol. Chem. 286: 23022–23030.

Yang, Q., Y.L. Zheng and C.C. Harris. 2005. POT1 and TRF2 cooperate to maintain telomeric integrity. Mol. Cell. Biol. 25: 1070–80.

Ye, J.Z., J.R. Donigian, M. van Overbeek, D. Loayza, Y. Luo, A.N. Krutchinsky et al. 2004. TIN2 binds TRF1 and TRF2 simultaneously and stabilizes the TRF2 complex on telomeres. J. Biol. Chem. 279: 47264–47271.

Yeeles, J.T. and M.S. Dillingham. 2010. The processing of double-stranded DNA breaks for recombinational repair by helicase-nuclease complexes. DNA Repair. 9: 276–285.

Yu, Z., J. Chen, B.N. Ford, M.E. Brackley and B.W. Glickman. 1999. Human DNA repair systems: an overview. Environ. Mol. Mutagen. 33: 3–20.

Zharkov, D.O. 2008. Base excision DNA repair. Cell. Mol. Life Sci. 65: 1544–1565.

Chapter 2

Telomerase Activity

Daniel Eduardo Gómez,[1,*] *Diego Luis Mengual Gómez*,[1]
Lina Maloukh[2] and *Romina Gabriela Armando*[1]

INTRODUCTION

Since its discovery in 1984 by Carol W. Greider and Elizabeth Blackburn in the ciliate *Tetrahymena*, telomerase has been object of numerous and extensive studies. This chapter focuses on the structure and function of the telomere/telomerase complex, its regulation, the non telomeric activities of telomerase and diseases caused by alteration in the telomere/telomerase complex.

TELOMERASE STRUCTURE AND MECHANISM OF TELOMERIC EXTENSION

A fundamental telomeric function is to act as a kind of buffer to support the erosion of DNA chromosomic ends due to the problem of terminal replication. Conventional DNA polymerases are unidirectional and they cannot copy all bases in the 3' end after primer removal. The result is that in each replication cycle, a given end of the chromosome cannot be synthesized completely and is lost. If organisms could not solve this replicative problem they could not pass their genetic charge completely from generation to generation. As a consequence, all the species should have, at least at the level of germinal cells, a mechanism that prevents the incomplete replication of the genomes. Different organisms have acquired evolutively different methods to prevent the lost of DNA from the ends of their chromosomes; however, most

[1] Laboratorio de Oncología Molecular, Departamento de Ciencia y Tecnología, Universidad Nacional de Quilmes, R. Sáenz Peña 352, Bernal B1876BXD, Buenos Aires, Argentina.
Email: dmengualgomez@gmail.com
Email: rgarmando@yahoo.com.ar
[2] University of Modern Sciences. Dubai. United Arab Emirates. Al Twar 3, 7A Street P.O. Box: 231931 Dubai. UAE.
Email: l.maloukh@ums.ae
* Corresponding author: degomez@unq.edu.ar

of the mammals use a specialized retrotranscriptase called telomerase (Greider and Blackburn 1985). The human holoenzyme telomerase is a ribonucleoprotein composed by a catalytic subunit, hTERT found in the cytoplasm and an RNA component (hTR) which acts as a template for the addition of a short repetitive sequence (dTTAGGG)n in the 3' end of the telomeric DNA found in the nucleus and species-specific accessory proteins (Wyatt et al. 2010). These accessory proteins regulate telomerase biogenesis, subcellular localization and function *in vivo*. For instance, analysis of affinity-purified telomerase from HeLa cells has identified protein components of human telomerase: dyskerin, nucleolar protein 10 (NOP10), non-histone protein 2 (NHP2), glycine-arginine rich 1 (GAR1), pontin/reptin, heat shock protein 90 (HSP90), p23 and TCAB1 (telomerase CB protein1) (Schmidt and Cech 2015). hTERT is synthesized in the cytoplasm and associates with the chaperones HSP90 and p23. hTR cotranscriptionally binds dyskerin, NOP10, NHP2 and the chaperone NAF1 (nuclear assembly factor 1 ribonucleoprotein), with NAF1 subsequently being replaced by GAR1 upon entry to CBs. Assembly of hTR and hTERT into catalytically active telomerase is aided by the ATPases Reptin and Pontin (Venteicher et al. 2008). Telomerase is recruited to Cajal Bodies (CBs) which are places of ensambly by its interaction with TCAB1. Thus, TCAB1 controls telomerase trafficking. In S phase of the cell cycle, telomerase is recruited to telomeres out of the CBs by the interaction of the TEN domain of hTERT with TPP1 and it is disassembled probably during M phase (Wojtyla et al. 2011, Schmidt and Cech 2015) (Fig. 1). Dyskerin, NHP2 and NOP10 are required for the stability and accumulation of human telomerase RNA (hTR) *in vivo* (Egan and Collins 2012). Besides its association with telomerase, dyskerin is a highly conserved nucleolar protein that, as part of a specialized nucleolar RNP, catalyses the pseudouridylation of specific residues in newly synthesized ribosomal RNAs and spliceosomal snRNA. Pontin and reptin have multiple roles. Both ATPases are associated with several chromatin remodeling complexes and have many functions including transcriptional regulation, DNA damage repair and telomerase activity. They can also interact with major oncogenic actors such as β-catenin and c-myc and regulate their oncogenic function. In humans the RNA component has an extension of 451 nucleotides of length and contains a sequence of 11 bp (5-CUAACCCUAAC-3') that encodes for telomeric repeats (Morin 1989). The gen hTR is present in the human genome also as a single copy in chromosome 3q26. Phylogenetic comparative analysis of vertebrate TR predicts three conserved domains: (i) the pseudoknot/template core domain, (ii) the CR4/CR5 domain and (iii) a box H/ACA domain (Chen et al. 2000). The core domain is essential for telomerase activity *in vitro* and *in vivo*. The CR4/CR5 is required for telomerase activity but is not essential for hTR stability in the cell and the box H/ACA domain is essential for TR stability, processing, nuclear localization and telomerase activity *in vivo* (Theimer and Feigon 2006).

Telomeric Elongation by Telomerase

Described as the first and most important function of telomerase, numerous reports have described this activity in detail (Nandakumar and Cech 2013). Briefly, elongation of the telomere by telomerase is a process that happens in different

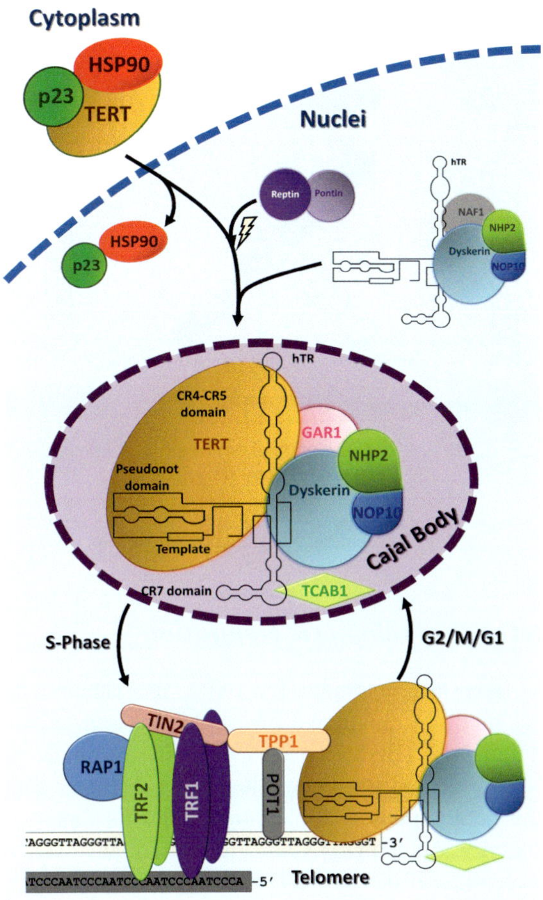

Fig. 1. Telomerase Structure and Assembly. hTERT is synthesized in the cytoplasm and associates with the chaperones HSP90 and p23. hTR cotranscriptionally binds dyskerin, NOP10, NHP2 and the chaperone NAF1, with NAF1 subsequently being replaced by GAR1 upon entry to CBs. Assembly of hTR and hTERT into catalytically active telomerase is aided by the ATPases Reptin and Pontin. Telomerase is recruited to Cajal Bodies (CBs) which are places of ensambly by its interaction with TCAB1, which controls telomerase trafficking. In S phase of the cell cycle, telomerase is recruited to telomeres out of the CBs by the interaction of the TEN domain of hTERT with TPP1 and it is disassembled probably during M phase.

stages. First, the nucleotides of the 3' extreme of the telomeric DNA are hybridized to the end of the RNA template, inside the RNA domain of the telomerase complex. The template sequence of 11 nucleotides is complementary to almost two telomeric repeats. Secondly, the gap in the extreme of the template is completed by synthesis, using triphosphate nucleotides in the catalytic site of the enzyme (hTERT). In this way, a complete hexanucleotidic repeat is assembled in the template. Finally, the synthesized strand is translocated in 5' direction in order to allow the formation of a new gap and the repetition of the cycle (Fig. 2).

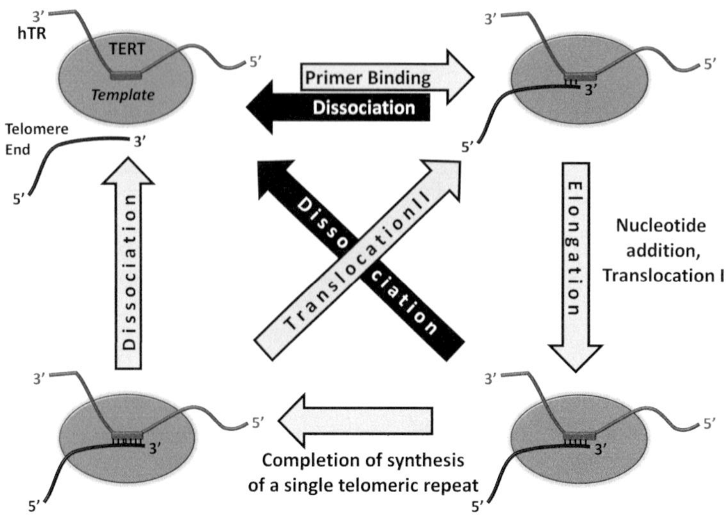

Fig. 2. Telomeric elongation by telomerase.

Alternative methods of telomeric elongation

Human cell lines that lack telomerase activity, are capable of maintaining or elongating their telomeres by alternative means, this has been called ALT (alternative lengthening of telomeres) (Bryan et al. 1997). In mammals that exhibit ALT, sequences of DNA are copied from one telomere to the other suggesting that ALT would involve the homologous recombination (HR) machinery. Critical changes occur in ALT telomeric chromatin to bypass the shelterin mediated repression of HR. Orphan nuclear receptors of the NR2C/F class have been found at ALT telomeres (Dejardin and Kingston 2009). Later, it was demonstrated that the nuclear receptors bind to a variant telomeric repeat sequence (5' GGGTCA 3') instead of the canonical repeat. This variant repeat is scarce in normal telomeres but they accumulate at ALT telomeres (Conomos et al. 2013). Recently Marzec et al. identified tandem 5' GGGTCA 3' repeats accumulating at ALT telomeres bound by NRC2C/F nuclear receptors (Marzec et al. 2015). Receptor binding to telomeres induced telomere cluster formation, which is required for HR in ALT. Although nuclear receptor binding to telomeres is very important to trigger ALT, a number of distinct events are required also for ALT demonstrating the complexity of this alternative way of telomeric elongation (Aeby and Lingner 2015).

Telomerase regulation

Telomerase activity contains multiple factors that allow its regulation, including the expression of TERT levels, post-transcriptional modifications, transportation and location and finally conformation and interactions with accessory proteins during assembly of the telomerase complex (Fig. 3). Telomerase levels are regulated at every

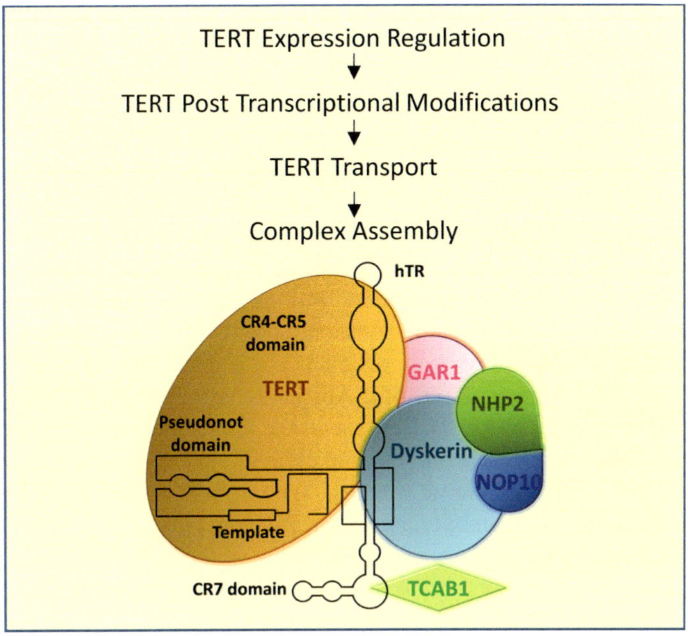

Fig. 3. Levels of telomerase regulation.

step of protein and RNA processing, as well as at the level of complex assembly and subcellular localization (Kelleher et al. 2002). Telomerase activity at the telomere is also regulated at the level of telomerase recruitment to the telomere. While the exact mechanism of telomerase recruitment is still not fully known, it is likely part of a negative feedback loop created by shelterin complex, a group of 6 proteins bound at the telomere that protect telomeres from DNA repair mechanisms and serve as negative regulators of telomerase extension of telomeres. The six proteins are Telomere Repeat Factor 1 (TRF1), Telomere Repeat Factor 2 (TRF2), Protection of Telomere 1 (POT1), Repressor/Activator Protein 1 (RAP1), TRF1- and TRF2-Interacting Nuclear Protein 2 (TIN2), and Tripeptidyl-peptidase 1 (TPP1) (Fig. 1). They can operate in smaller subsets to regulate the telomeric length or its protection (De Boeck et al. 2009). The prevailing model of shelterin mediated telomere length regulation is that longer telomeres recruit more shelterin complex, which in turn limits future telomerase elongation. This negative feedback loop is thought to be responsible for the stable telomere length found in cancer cells, and is likely responsible for at least partially maintaining telomere length homeostasis in germ cells and other stem-like cells in which telomerase is active (Forsyth et al. 2002). Thus, it would be possible that other mechanisms regulating telomerase recruitment or processivity exist depending on whether cells are at equilibrium conditions versus non-equilibrium conditions.

As already reported, TERT splice variants may be expressed in normal, pre-crisis and alternative lengthening of telomeres cells (ALT) that lack detectable telomerase activity (Kilian et al. 1997, Ulaner et al. 1998, Ulaner et al. 2000). Thus, transcriptional control of TERT is supposed to play a crucial role in the complex regulation of telomerase activity. Post-translational regulation of telomerase activity can occur via reversible phosphorylation of TERT catalytic subunit at specific serine/threonine or tyrosine residues. Due to multiple kinase and phosphatase activators and inhibitors the telomerase phosphorylation status may affect its structure, localization and enzyme activity (Cong et al. 2002). Numerous non-specific phosphorylation sites within TERT protein are postulated but only a few of them appear to be the key residues, and their phosphorylation influences telomerase activity (activation and inhibition). Specific phosphorylation site at TERT is present at the proline rich region. It was revealed that the contribution of c-Abl tyrosine kinase to TERT phosphorylation at specific tyrosine residues led to decreased telomerase activity. It was shown that overexpression of c-Abl inhibited cell growth by causing cell cycle arrest (Sawyers et al. 1994). Because of the role of c-Abl in stress response to DNA damage, exposure of cells to ionizing radiation led to a significant increase in TERT phosphorylation by c-Abl. It was also demonstrated that c-Abl phosphorylated TERT is leading to inhibition of telomerase activity and decrease in telomere length (Kharbanda et al. 2000) suggesting a direct association between c-Abl and TERT. In conclusion, TERT expression is regulated at both, transcriptional and post-transcriptional levels, with the alternative splicing of TERT also involved in the control of telomerase activity. However, contradictive reports concern the correlation of telomere length with telomerase activity or TERT expression in different cells which might confirm the tissue-specificity of those regulatory mechanisms.

As described above, translocation of TR and TERT is regulated and multiple nuclear structures participate in transport and biogenesis of telomerase (Tomlinson et al. 2006). Throughout most of the cell cycle TR is present in CBs that act as its transmitters to telomeres; these subnuclear structures are general sites of RNP assembly and RNA modification (Cioce and Lamond 2005, Jady et al. 2006). In contrary to TR, TERT is located in distinct nucleoplasmic foci and therefore, the two main subunits of telomerase are separated during almost the whole cell cycle. In early S phase TERT is translocated to nucleoli. At the same time CBs containing TR accumulate at the periphery of nucleoli. TR accumulates at the pole of CBs that precedes localization to telomeres in mid-S phase when CBs deliver telomerase to individual telomeres. Furthermore, it was revealed that the same kinases and phosphatases that act during S-phase may modify telomerase subunits. However, the mechanisms involved in targeting and accumulation of TR are not fully understood. To date, within telomerase RNA molecule the CAB box and H/ACA motif has been identified to influence the TR translocation to CBs and nucleoli (Lukowiak et al. 2001, Jady et al. 2006). In the same way, one of TERT domains is known to mediate nucleolar translocation (Etheridge et al. 2002, Yang et al. 2002).

Assembly of telomerase complex

Prevention of premature binding of the essential telomerase subunits (TERT and TR) is possible due to different sites of their compartmentalization and keeping them away from their substrates (telomeres). Thus, two telomerase assembling sites are possible during S phase: at the telomere ends or in CBs. It has been suggested that survival of motor neuron (SMN) complex, an RNP assembly factor present in CBs, takes part in telomerase biogenesis. It was demonstrated that TR is associated with GAR1, a protein which interacts with SMN complex. Also, TERT is supposed to indirectly influence the trafficking of TR or a transient interaction of the two components that contribute to TR localization (Kellermann et al. 2015).

Certain mutation in telomerase proteins leading to pathologies, that will be explained later, could intervene in telomerase regulation. Numerous unique mutations have been identified within the DKC1 gene, encoding for dyskerin (Cossu et al. 2002, Vulliamy et al. 2008). Mutations in a limited number of families have also been reported in NOLA2, encoding for NHP2 (Trahan et al. 2010); and NOLA3, encoding for NOP10 (Walne et al. 2007). These mutations retain wild-type telomerase activity *in vitro* while reducing the amount of active telomerase within the cell. Since the discovery of TCAB1 and its gene, WRD79, there has been a report of mutations linked to two cases of autosomal recessive DC. These mutations produce defects in TR trafficking, reducing the amount of active enzyme (Zhong et al. 2011).

Regulation of telomerase by telomere binding

Interaction of telomerase with numerous telomere binding proteins (TBP) that may influence telomerase enzyme activity is considered another form of telomerase activity regulation. It is supposed that binding some of them to telomeres, therefore making it impossible for telomerase to access the chromosome ends, is an indirect way of regulation. Three-state model of telomere length regulation were studied (Wang et al. 2007). Firstly, POT1 is directly bound at the 3' end of telomere and associates with TPP1. This POT1-TPP1 position prevents binding of telomerase to chromosome ends. Secondly, TPP1-POT1 association enhanced POT1 affinity for telomeric sDNA and TPP1 associates with the telomerase, providing a physical link between telomerase and the shelterin complex (Xin et al. 2007). According to this model, in the next state these proteins are removed from their binding sites by an unidentified mechanism. Postranslational modification or disruption of shelterin might be involved in the process. Thirstly, released POT1-TPP1 complex may serve as an activator of telomerase during telomere extension. When elongated, telomere reaches a certain threshold, the newly synthesized repeats bind shelterin complexes and the 3' end of the overhang is re-bound by POT1-TPP1. This causes telomerase inhibition and return of the telomere to the first state of the complex (Xin et al. 2007). TRF1 and TRF2 are the main proteins responsible for telomerase negative feedback control in mammals. They are bound to double stranded DNA at T-loop, which is a 'closed' state of telomere, which telomerase cannot access and therefore extends the telomere terminus. TRF1 and TRF2 act as negative regulators of telomere length because they are involved in T-loop formation (Smogorzewska et al. 2000, Shore and Bianchi 2009). TRF1 and TRF2 were shown to act in *cis* to repress telomere elongation. TRF1 was reported to repress telomerase action on telomeres while, on the contrary, TRF2 appears to activate a telomeric degradation without showing any influence on telomerase (Ancelin et al. 2002). Other proteins with negative-feedback regulation of telomere length have been identified in human cells. The proteins acting on TRF1 are tankyrase 1 and 2 (TANK 1 and 2), TIN2, PINX1, three TRF1-interacting factors but also hRAP1 which interacts with TRF2 (Sinha and Bandyopadhyay 2011). PINX1 can inhibit telomerase by forming a stable complex with catalytic subunit of telomerase and TRF1 molecule. It binds with TERT by its telomerase inhibitory domain (TID) placed at C terminal 74 aa. The human repressor activator protein 1 (hRap1) was identified as a protein that specifically interacts with TRF2 and negatively regulates telomere length *in vivo*.

Recently, it was found that heregulin is a new interactor of the shelterin complex in human telomerase. Specially, the isoform β2 of the heregulin family of growth factors is able to interact with RAP1 and with TRF2. Deletion analysis of the mentioned isoform confirmed that a putative nuclear localization signal (NLS) was necessary for the nuclear isoform to exert a negative regulation of telomere length whereas the N-terminus (extracellular) amino acids of it were sufficient to interact with RAP1/TRF2 and promote telomere shortening (Menendez et al. 2015).

Also Polo-like kinase 1 (Plk1) was identified as a new telomerase-associated protein. Plk1 can interact with hTERT independently of its kinase activity. More importantly, it was found that Plk1 is associated with active telomerase complex. In addition, the knockdown of Plk1 caused the reduction of the telomerase activity, whereas its overexpression increased telomerase activity. Also Plk1 affected hTERT stability by inhibiting its ubiquitin-mediated degradation. These observations suggest that Plk1 is a positive modulator of telomerase by enhancing the stability of hTERT (Huang et al. 2015). Also it seems that protochadherin 10 (PCD10) may play a negative regulating telomerase activity (Zhou et al. 2015).

Pharmacological drugs could also affect telomerase homeostasis/function. A complete description of the drugs is out of the scope of this chapter, however just as an example, bortezomib in certain cell types led to significant inhibition of hTERT and telomerase expression, widespread dysregulation of shelterin protein expression, and telomere shortening, thereby triggering telomere dysfunction and DNA damage (Ci et al. 2015).

Non telomeric activities of telomerase

Telomerase is essential for the long-term proliferation potential of stem cells and cancer cells, and for normal tissue renewal. However, other functions have been described beyond its action at the telomeric level. Indeed, TERT can function as a transcriptional modulator of the Wnt-β-catenin signaling pathway (Park et al. 2009). TERT functions as a cofactor in a β-catenin transcriptional complex through interactions with BRG1, which is an SWI/SNF related chromatin remodeling protein, allowing regulation of target genes of this pathway (Jaitner et al. 2012). Good examples are Cyc-D1, which plays a key role in cell cycle progression (Tang et al. 2015) and c-Myc with a direct link to cell growth, differentiation, and apoptosis (Park et al. 2009). In addition, TERT in a complex with RMRP can act as an RNA-dependent RNA polymerase (Maida et al. 2009). The TERT-RMRP complex acts as an RDRP and processes RMRP into double-stranded RNA (dsRNA), which is then processed by the endoribonuclease Dicer into small interfering RNA (siRNA), which controls RMRP endogenous levels. Thus, TERT-RMRP-RDRP regulates RMRP levels by a negative-feedback control mechanism. A role for telomerase in the regulation of apoptosis in a telomere maintenance-independent manner has been found (Maida et al. 2009). TERT contains a mitochondrial localization signal peptide at its N-terminal that targets TERT to mitochondria where it is active (Santos et al. 2004). Furthermore, it was shown that telomerase sensitizes mitochondrial DNA to hydrogen peroxide-induced oxidative damage, probably through the modulation of metal homeostasis (Santos et al. 2004). The mitochondrial localization of telomerase also has an important role in apoptosis (Cong and Shay 2008).

Telomeropaties

An enormous and conclusive amount of research have proved the role that telomerase plays in cancer (reviewed by (Gomez et al. 2012)), however a constellation of related genetic diseases are caused by defects in the telomere maintenance machinery. These disorders, often referred as telomeropathies, share symptoms and molecular mechanisms, and mounting evidence indicates they are points along a spectrum of disease (Holohan et al. 2014). The first telomeropathy described and the best known is a disease-associated with mutations in human dyskerin, identified in patients afflicted with a rare, multi-system disorder called dyskeratosis congenita (DKC) (Walne and Dokal 2008). More recently, telomerase mutations have been detected in the context of aplastic anemia (Vulliamy et al. 2002), Hoyeraal Hreidarsson syndrome (HHS) (Nishio and Kojima 2010) idiopathic pulmonary fibrosis (Armanios 2012) Revesz syndrome (Linnankivi et al. 2013) and Coat Plus Syndrome (Anderson et al. 2012). The clinical manifestations of dyskeratosis congenita generally appear during childhood and include a monocutaneous triad of abnormal skin pigmentation, nail dystrophy and oral leukoplasia. The symptoms are accompanied by a spectrum of other somatic abnormalities such as developmental delay, premature hair loss and organ failure. Bone marrow failure is the principal cause of premature mortality. Dyskerin protein is encoded by a gene called DKC1 which is located on the long (q) arm of the X chromosome at position 28 (Xq28). DKC1 gene comprises 15 exons spanning at least 16 kb and is transcribed into a widely expressed 2.6-kb (Drachtman and Alter 1995). Mutations found within DKC1 gene has been linked to an x-linked recessive form of the inherited disorder disease dyskeratosis congenita (DKC) (Vulliamy et al. 2001). There is also an autosomal dominant form of DKC disease that is linked to mutations either in the hTR or TERT (Vulliamy and Dokal 2008). The x-linked recessive form of DKC has more severe clinical presentation than the autosomal dominant form. Those mutations can cause amino acid substitutions, deletions and the loss of the entire exon 15 within dyskerin. Some examples of those mutations are r.54_57delaacu (in the 5'UTR domain and the Template region), r.100u > a (in the Pseudoknot domain and the P2b flanking region) and c.1142G > C (in the N-Terminal) (Sirinavin and Trowbridge 1975).

HHS is a severe form of DKC, with the addition of intrauterine growth retardation, bone marrow failure, immunodeficiency, cerebellar hypoplasia, and microcephaly. HHS patients present very short telomeres and a high mortality rate (Walne et al. 2013).

Revesz syndrome is a very rare disease, characterized by symptoms of HHS with the addition of exudative retinopathy (Revesz et al. 1992).

Coat Plus syndrome may exhibit symptoms of HHS and Revesz syndrome, with the addition of cerebral calcifications (Riyaz et al. 2007). Presently it is difficult to distinguish HHS, Revesz and Coat Plus syndromes due to the incomplete penetrance of the symptoms and the very low number of cases. Because of the overlapping symptoms and causes of these disorders, it is likely that they represent a single disease entity with multiple genetic mechanisms for the same pathological conditions and not three distinct disorders.

Aplastic anemia is a hematological disorder characterized by reduced red blood cell counts, bone marrow and liver failure and lung disease.

Idiopathic pulmonary fibrosis is a chronic, progressive, and fatal disease that is defined by progressive failure of the lung coincident with fibrosis and inflammation (Meltzer and Noble 2008). The unifying molecular characteristic of these diseases is that patients harbor telomeres that are significantly shorter than age-matched control subjects (Armanios 2009).

In addition to classically linked disorders of impaired telomere maintenance, a number of diseases not typically associated with telomere biology have been reported, which includes dysfunction at the telomeres. We refer to them as secondary telomeropathies because it is not known yet if the reported telomere dysfunction is the cause or just one of the effects of the diseases. They include Fanconi Anemia (Joksic et al. 2012), and ICF (immunodeficiency, centromeric region instability and facial anomalies syndrome type I) (Yehezkel et al. 2013), Progeria, Hutchinson-Gilford Progeria Syndrome, Werner's syndrome, Rothmund-Thomson syndrome and Cockayne syndrome (the last four are rare premature aging disorders) (Hicks et al. 2007, Decker et al. 2009, Batenburg et al. 2012, Edwards et al. 2014). As described, there are a number of well-known diseases caused by defects in telomere maintenance. Furthermore, we listed a number of related syndromes that may be unrecognized telomeropathies (secondary telomeropathies). Probably, in the future new syndromes will be revealed having a relationship with telomere dysfunction, describing the wide spectrum that these diseases represent.

Conclusions

Although great advances has been done about telomerase in a short period of time, the complexity of its structure, regulation, new functions non related with the telomere and the discovery of many diseases related directly or indirectly to telomerase, guarantee that the field will still bring new and important discoveries that could partially rewrite some aspects of cell biology.

Acknowledgments

This study was supported by grants from UNQ and ANPCyT (Argentina). Daniel E. Gomez and Diego Mengual Gómez are members of the National Research Council (CONICET, Argentina).

Keywords: Telomere, extension, elongation, telomerase regulation

References

Aeby, E. and J. Lingner. 2015. ALT telomeres get together with nuclear receptors. Cell. 160(5): 811–813.

Ancelin, K., M. Brunori, S. Bauwens, C.E. Koering, C. Brun, M. Ricoul et al. 2002. Targeting assay to study the cis functions of human telomeric proteins: evidence for inhibition of telomerase by TRF1 and for activation of telomere degradation by TRF2. Mol. Cell Biol. 22(10): 3474–3487.

Anderson, B.H., P.R. Kasher, J. Mayer, M. Szynkiewicz, E.M. Jenkinson, S.S. Bhaskar et al. 2012. Mutations in CTC1, encoding conserved telomere maintenance component 1, cause Coats plus. Nat. Genet. 44(3): 338–342.

Armanios, M. 2009. Syndromes of telomere shortening. Annu. Rev. Genomics Hum. Genet. 10: 45–61.

Armanios, M. 2012. Telomerase and idiopathic pulmonary fibrosis. Mutat. Res. 730(1-2): 52–58.

Batenburg, N.L., T.R. Mitchell, D.M. Leach, A.J. Rainbow and X.D. Zhu. 2012. Cockayne syndrome group B protein interacts with TRF2 and regulates telomere length and stability. Nucleic Acids Res. 40(19): 9661–9674.

Bryan, T.M., A. Englezou, L. Dalla-Pozza, M.A. Dunham and R.R. Reddel. 1997. Evidence for an alternative mechanism for maintaining telomere length in human tumors and tumor-derived cell lines. Nat. Med. 3(11): 1271–1274.

Ci, X., B. Li, X. Ma, F. Kong, C. Zheng, M. Bjorkholm et al. 2015. Bortezomib-mediated down-regulation of telomerase and disruption of telomere homeostasis contributes to apoptosis of malignant cells. Oncotarget. 6(35): 38079–38092.

Cioce, M. and A.I. Lamond. 2005. Cajal bodies: a long history of discovery. Annu. Rev. Cell Dev. Biol. 21: 105–131.

Cong, Y. and J.W. Shay. 2008. Actions of human telomerase beyond telomeres. Cell Res. 18(7): 725–732.

Cong, Y.S., W.E. Wright and J.W. Shay. 2002. Human telomerase and its regulation. Microbiol. Mol. Biol. Rev. 66(3): 407–425.

Conomos, D., H.A. Pickett and R.R. Reddel. 2013. Alternative lengthening of telomeres: remodeling the telomere architecture. Front Oncol. 3: 27.

Cossu, F., T.J. Vulliamy, A. Marrone, M. Badiali, A. Cao and I. Dokal. 2002. A novel DKC1 mutation, severe combined immunodeficiency (T+B-NK-SCID) and bone marrow transplantation in an infant with Hoyeraal-Hreidarsson syndrome. Br. J. Haematol. 119(3): 765–768.

Chen, J.L., M.A. Blasco and C.W. Greider. 2000. Secondary structure of vertebrate telomerase RNA. Cell. 100(5): 503–514.

De Boeck, G., R.G. Forsyth, M. Praet and P.C. Hogendoorn. 2009. Telomere-associated proteins: cross-talk between telomere maintenance and telomere-lengthening mechanisms. J. Pathol. 217(3): 327–344.

Decker, M.L., E. Chavez, I. Vulto and P.M. Lansdorp. 2009. Telomere length in Hutchinson-Gilford progeria syndrome. Mech. Ageing Dev. 130(6): 377–383.

Dejardin, J. and R.E. Kingston. 2009. Purification of proteins associated with specific genomic Loci. Cell. 136(1): 175–186.

Drachtman, R.A. and B.P. Alter. 1995. Dyskeratosis congenita. Dermatol. Clin. 13(1): 33–39.

Edwards, D.N., D.K. Orren and A. Machwe. 2014. Strand exchange of telomeric DNA catalyzed by the Werner syndrome protein (WRN) is specifically stimulated by TRF2. Nucleic Acids Res. 42(12): 7748–7761.

Egan, E.D. and K. Collins. 2012. Biogenesis of telomerase ribonucleoproteins. RNA. 18(10): 1747–1759.

Etheridge, K.T., S.S. Banik, B.N. Armbruster, Y. Zhu, R.M. Terns, M.P. Terns et al. 2002. The nucleolar localization domain of the catalytic subunit of human telomerase. J. Biol. Chem. 277(27): 24764–24770.

Forsyth, N.R., W.E. Wright and J.W. Shay. 2002. Telomerase and differentiation in multicellular organisms: turn it off, turn it on, and turn it off again. Differentiation. 69(4-5): 188–197.

Gomez, D.E., R.G. Armando, H.G. Farina, P.L. Menna, C.S. Cerrudo, P.D. Ghiringhelli et al. 2012. Telomere structure and telomerase in health and disease (review). Int. J. Oncol. 41(5): 1561–1569.

Greider, C.W. and E.H. Blackburn. 1985. Identification of a specific telomere terminal transferase activity in Tetrahymena extracts. Cell. 43(2 Pt 1): 405–413.

Hicks, M.J., J.R. Roth, C.A. Kozinetz and L.L. Wang. 2007. Clinicopathologic features of osteosarcoma in patients with Rothmund-Thomson syndrome. J. Clin. Oncol. 25(4): 370–375.

Holohan, B., W.E. Wright and J.W. Shay. 2014. Cell biology of disease: Telomeropathies: an emerging spectrum disorder. J. Cell Biol. 205(3): 289–299.

Huang, Y., L. Sun, N. Liu, Q. Wei, L. Jiang, X. Tong et al. 2015. Polo-like Kinase 1 (Plk1) Up-regulates telomerase activity by affecting human telomerase reverse transcriptase (hTERT) stability. J. Biol. Chem. 290(30): 18865–18873.

Jady, B.E., P. Richard, E. Bertrand and T. Kiss. 2006. Cell cycle-dependent recruitment of telomerase RNA and Cajal bodies to human telomeres. Mol. Biol. Cell. 17(2): 944–954.

Jaitner, S., J.A. Reiche, A.J. Schaffauer, E. Hiendlmeyer, H. Herbst, T. Brabletz et al. 2012. Human telomerase reverse transcriptase (hTERT) is a target gene of beta-catenin in human colorectal tumors. Cell Cycle. 11(17): 3331–3338.

Joksic, I., D. Vujic, M. Guc-Scekic, A. Leskovac, S. Petrovic, M. Ojani et al. 2012. Dysfunctional telomeres in primary cells from Fanconi anemia FANCD2 patients. Genome Integr. 3(1): 6.

Kelleher, C., M.T. Teixeira, K. Forstemann and J. Lingner. 2002. Telomerase: biochemical considerations for enzyme and substrate. Trends Biochem. Sci. 27(11): 572–579.

Kellermann, G., M. Kaiser, F. Dingli, O. Lahuna, D. Naud-Martin, F. Mahuteau-Betzer et al. 2015. Identification of human telomerase assembly inhibitors enabled by a novel method to produce hTERT. Nucleic Acids Res. 43(15): e99.

Kharbanda, S., V. Kumar, S. Dhar, P. Pandey, C. Chen, P. Majumder et al. 2000. Regulation of the hTERT telomerase catalytic subunit by the c-Abl tyrosine kinase. Curr. Biol. 10(10): 568–575.

Kilian, A., D.D. Bowtell, H.E. Abud, G.R. Hime, D.J. Venter, P.K. Keese et al. 1997. Isolation of a candidate human telomerase catalytic subunit gene, which reveals complex splicing patterns in different cell types. Hum. Mol. Genet. 6(12): 2011–2019.

Linnankivi, T., A. Polvi, O. Makitie, A.E. Lehesjoki and T. Kivela. 2013. Cerebroretinal microangiopathy with calcifications and cysts, Revesz syndrome and aplastic anemia. Bone Marrow Transplant. 48(1): 153.

Lukowiak, A.A., A. Narayanan, Z.H. Li, R.M. Terns and M.P. Terns. 2001. The snoRNA domain of vertebrate telomerase RNA functions to localize the RNA within the nucleus. RNA. 7(12): 1833–1844.

Maida, Y., M. Yasukawa, M. Furuuchi, T. Lassmann, R. Possemato, N. Okamoto et al. 2009. An RNA-dependent RNA polymerase formed by TERT and the RMRP RNA. Nature. 461(7261): 230–235.

Marzec, P., C. Armenise, G. Perot, F.M. Roumelioti, E. Basyuk, S. Gagos et al. 2015. Nuclear-receptor-mediated telomere insertion leads to genome instability in ALT cancers. Cell. 160(5): 913–927.

Meltzer, E.B. and P.W. Noble. 2008. Idiopathic pulmonary fibrosis. Orphanet. J. Rare. Dis. 3: 8.

Menendez, J.A., M.A. Rubio, J. Campisi and R. Lupu. 2015. Heregulin, a new regulator of telomere length in human cells. Oncotarget. 6(37): 39422–39436.

Morin, G.B. 1989. The human telomere terminal transferase enzyme is a ribonucleoprotein that synthesizes TTAGGG repeats. Cell. 59(3): 521–529.

Nandakumar, J. and T.R. Cech. 2013. Finding the end: recruitment of telomerase to telomeres. Nat. Rev. Mol. Cell Biol. 14(2): 69–82.

Nishio, N. and S. Kojima. 2010. Recent progress in dyskeratosis congenita. Int. J. Hematol. 92(3): 419–424.

Park, J.I., A.S. Venteicher, J.Y. Hong, J. Choi, S. Jun, M. Shkreli et al. 2009. Telomerase modulates Wnt signalling by association with target gene chromatin. Nature. 460(7251): 66–72.

Park, S.E., H.S. Yoo, C.Y. Jin, S.H. Hong, Y.W. Lee, B.W. Kim et al. 2009. Induction of apoptosis and inhibition of telomerase activity in human lung carcinoma cells by the water extract of Cordyceps militaris. Food Chem. Toxicol. 47(7): 1667–1675.

Revesz, T., S. Fletcher, L.I. al-Gazali and P. DeBuse. 1992. Bilateral retinopathy, aplastic anaemia, and central nervous system abnormalities: a new syndrome? J. Med. Genet. 29(9): 673–675.

Riyaz, A., N. Riyaz, M.P. Jayakrishnan, P.T. Mohamed Shiras, V.T. Ajith Kumar and B.S. Ajith. 2007. Revesz syndrome. Indian J. Pediatr. 74(9): 862–863.

Santos, J.H., J.N. Meyer, M. Skorvaga, L.A. Annab and B. Van Houten. 2004. Mitochondrial hTERT exacerbates free-radical-mediated mtDNA damage. Aging Cell. 3(6): 399–411.

Sawyers, C.L., J. McLaughlin, A. Goga, M. Havlik and O. Witte. 1994. The nuclear tyrosine kinase c-Abl negatively regulates cell growth. Cell. 77(1): 121–131.

Schmidt, J.C. and T.R. Cech. 2015. Human telomerase: biogenesis, trafficking, recruitment, and activation. Genes Dev. 29(11): 1095–1105.

Shore, D. and A. Bianchi. 2009. Telomere length regulation: coupling DNA end processing to feedback regulation of telomerase. EMBO J. 28(16): 2309–2322.

Sinha, S.K. and S. Bandyopadhyay. 2011. Conformational fluctuations of a protein-DNA complex and the structure and ordering of water around it. J. Chem. Phys. 135(24): 245104.

Sirinavin, C. and A.A. Trowbridge. 1975. Dyskeratosis congenita: clinical features and genetic aspects. Report of a family and review of the literature. J. Med. Genet. 12(4): 339–354.

Smogorzewska, A., B. van Steensel, A. Bianchi, S. Oelmann, M.R. Schaefer, G. Schnapp et al. 2000. Control of human telomere length by TRF1 and TRF2. Mol. Cell Biol. 20(5): 1659–1668.

Tang, B., R. Xie, Y. Qin, Y.F. Xiao, X. Yong, L. Zheng et al. 2015. Human telomerase reverse transcriptase (hTERT) promotes gastric cancer invasion through cooperating with c-Myc to upregulate heparanase expression. Oncotarget.

Theimer, C.A. and J. Feigon. 2006. Structure and function of telomerase RNA. Curr. Opin. Struct. Biol. 16(3): 307–318.

Tomlinson, R.L., T.D. Ziegler, T. Supakorndej, R.M. Terns and M.P. Terns. 2006. Cell cycle-regulated trafficking of human telomerase to telomeres. Mol. Biol. Cell. 17(2): 955–965.

Trahan, C., C. Martel and F. Dragon. 2010. Effects of dyskeratosis congenita mutations in dyskerin, NHP2 and NOP10 on assembly of H/ACA pre-RNPs. Hum. Mol. Genet. 19(5): 825–836.

Ulaner, G.A., J.F. Hu, T.H. Vu, L.C. Giudice and A.R. Hoffman. 1998. Telomerase activity in human development is regulated by human telomerase reverse transcriptase (hTERT) transcription and by alternate splicing of hTERT transcripts. Cancer Res. 58(18): 4168–4172.

Ulaner, G.A., J.F. Hu, T.H. Vu, H. Oruganti, L.C. Giudice and A.R. Hoffman. 2000. Regulation of telomerase by alternate splicing of human telomerase reverse transcriptase (hTERT) in normal and neoplastic ovary, endometrium and myometrium. Int. J. Cancer. 85(3): 330–335.

Venteicher, A.S., Z. Meng, P.J. Mason, T.D. Veenstra and S.E. Artandi. 2008. Identification of ATPases pontin and reptin as telomerase components essential for holoenzyme assembly. Cell. 132(6): 945–957.

Vulliamy, T., A. Marrone, F. Goldman, A. Dearlove, M. Bessler, P.J. Mason et al. 2001. The RNA component of telomerase is mutated in autosomal dominant dyskeratosis congenita. Nature. 413(6854): 432–435.

Vulliamy, T., A. Marrone, I. Dokal and P.J. Mason. 2002. Association between aplastic anaemia and mutations in telomerase RNA. Lancet. 359(9324): 2168–2170.

Vulliamy, T., R. Beswick, M. Kirwan, A. Marrone, M. Digweed, A. Walne et al. 2008. Mutations in the telomerase component NHP2 cause the premature ageing syndrome dyskeratosis congenita. Proc. Natl. Acad. Sci. USA. 105(23): 8073–8078.

Vulliamy, T.J. and I. Dokal. 2008. Dyskeratosis congenita: the diverse clinical presentation of mutations in the telomerase complex. Biochimie. 90(1): 122–130.

Walne, A.J., T. Vulliamy, A. Marrone, R. Beswick, M. Kirwan, Y. Masunari et al. 2007. Genetic heterogeneity in autosomal recessive dyskeratosis congenita with one subtype due to mutations in the telomerase-associated protein NOP10. Hum. Mol. Genet. 16(13): 1619–1629.

Walne, A.J. and I. Dokal. 2008. Dyskeratosis Congenita: a historical perspective. Mech. Ageing Dev. 129(1-2): 48–59.

Walne, A.J., T. Vulliamy, M. Kirwan, V. Plagnol and I. Dokal. 2013. Constitutional mutations in RTEL1 cause severe dyskeratosis congenita. Am. J. Hum. Genet. 92(3): 448–453.

Wang, F., E.R. Podell, A.J. Zaug, Y. Yang, P. Baciu, T.R. Cech et al. 2007. The POT1-TPP1 telomere complex is a telomerase processivity factor. Nature. 445(7127): 506–510.

Wojtyla, A., M. Gladych and B. Rubis. 2011. Human telomerase activity regulation. Mol. Biol. Rep. 38(5): 3339–3349.

Wyatt, H.D., S.C. West and T.L. Beattie. 2010. InTERTpreting telomerase structure and function. Nucleic Acids Res. 38(17): 5609–5622.
Xin, H., D. Liu, M. Wan, A. Safari, H. Kim, W. Sun et al. 2007. TPP1 is a homologue of ciliate TEBP-beta and interacts with POT1 to recruit telomerase. Nature. 445(7127): 559–562.
Yang, Y., Y. Chen, C. Zhang, H. Huang and S.M. Weissman. 2002. Nucleolar localization of hTERT protein is associated with telomerase function. Exp. Cell Res. 277(2): 201–209.
Yehezkel, S., R. Shaked, S. Sagie, R. Berkovitz, H. Shachar-Bener, Y. Segev et al. 2013. Characterization and rescue of telomeric abnormalities in ICF syndrome type I fibroblasts. Front Oncol. 3: 35.
Zhong, F., S.A. Savage, M. Shkreli, N. Giri, L. Jessop, T. Myers et al. 2011. Disruption of telomerase trafficking by TCAB1 mutation causes dyskeratosis congenita. Genes Dev. 25(1): 11–16.
Zhou, L.N., X. Hua, W.Q. Deng, Q.N. Wu, H. Mei and B. Chen. 2015. PCDH10 interacts with hTERT and negatively regulates telomerase activity. Medicine (Baltimore). 94(50): e2230.

Chapter 3

Measurement of Telomere Length

Álvaro Pejenaute and *Guillermo Zalba Goñi**

INTRODUCTION

In recent years we have witnessed the growing interest that the measurement of the length of telomeres has awakened in the field of biomedical research and clinical practice as well as in various commercial applications. This requires not only the use of very accurate techniques but also high-throughput methods.

Telomeric DNA shortening is a physiological process associated to aging and it has also been linked to several age related diseases like tumorigenesis in cancer, metabolic diseases, Alzheimer and cardiovascular pathologies (coronary artery disease, heart failure and atherosclerosis). But it is not just the mean telomere length of a sample or even a cell, recently it has been highlighted the importance of critically short telomeres as a good indicator of telomere dysfunction, so both parameters have to be taken into account when studying telomere length.

The aim of this chapter is to review the main methods in telomere length measurement comparing the technologies and pointing their strong points and weaknesses. Actually these methodologies can be classified in four groups:

1. Terminal restriction fragment analysis: Based in DNA digestion by restriction endonucleases and its analysis by Southern blot.
2. Polymerase chain reaction (PCR) based methods: In which telomeres are amplified using specific primers by PCR.
3. Quantitative fluorescence *in situ* hybridization methods: Based on the hybridization of a fluorescent probe complementary to the telomeric DNA and its quantification by digital fluorescence microscopy.
4. Single telomere length analysis technology: A ligation mediated PCR method.

Finally two less used methods are described at the end of the chapter: Hybridization protection assay and Primed *in situ*.

Department of Biochemistry and Genetics, University of Navarra.
* Corresponding author: gzalba@unav.es

Terminal Restriction Fragment Analysis

Terminal restriction fragment (TRF) analysis was the first technique used to measure average telomere length, and since it was for many years the only method available it is described as the "gold standard" in telomere quantification (Montpetit et al. 2014).

It is important to take into account that genomic DNA extraction and its integrity is critical for TRF analysis (and also for the following methods), as sample degradation could lead to an underestimation of telomere length. The procedure begins with the digestion of genomic DNA with a combination of frequent cutting restriction endonucleases (in particular those that recognizes sequences of four bases), for example *Hinf*I/*Rsa*I or *Hph*I/*Mnl*I (Kimura et al. 2010). Due to its TTAGGG repeats sequence telomeric DNA lacks the presence of these enzymes target sites, and it is keep intact while chromosomal DNA is cut in small fragments of less than 800 bp.

Samples are loaded into an agarose gel and resolved, based on size, by electrophoresis with an appropriate molecular weight ladder. Uncut telomere fragments are then visualized using a labelled probe complementary to the telomeric sequence in the developed agarose gel (in-gel hybridization), or after transferring the DNA to a nylon membrane (Southern blot). Each sample will present a smear as the size distribution of all telomere fragments, so an average telomere length is calculated based on the intensity and size using this formula: $\sum(OD_i)/\sum(OD_i/L_i)$, where OD_i is optical density at position i and L_i is DNA length at position i (Kimura et al. 2010).

The principal advantage of TRF analysis is that provides an absolute quantification in kilobases of average telomere length per sample. Besides, it is a well-established method which does not require costly or specialized equipment, accurate and with relatively small coefficient of variation (< 2%) if experimental conditions (endonucleases cocktail selection, DNA quantity and quality...) are appropriate. These characteristics have made TRF analysis commonly used and the "gold standard" procedure, used to calibrate and validate other methods and allowing comparison between different studies performed by different laboratories.

One shortcoming of TRF analysis is that the telomere length quantification is overestimated as it includes a part of subtelomeric region (Fig. 1), which can also vary in length depending of the endonuclease cocktail used for the digestion of genomic DNA, thus adding inter-individual variation with independence of the real telomere length. This diversity of methodological approaches importantly underlies in the high variability of results found among different laboratories. Other drawback is the great amount of starting DNA (at least 1 to 3 micrograms) making TRF determination preferentially applicable to telomere length determination in blood samples rather than in other tissues. Also has the limitation of being labor intensive and time consuming, which makes it unsuitable to large epidemiological studies. Finally the main disadvantages of TRF are its relative insensitivity to the detection of very small telomeres and the fact that is not chromosome specific (as the result obtained is the mean telomere length of the whole sample). Very small telomeres have few TTAGGG tandem sequences and could not efficiently bind the labelled probe, since recent studies have highlighted the relevance of short telomeres

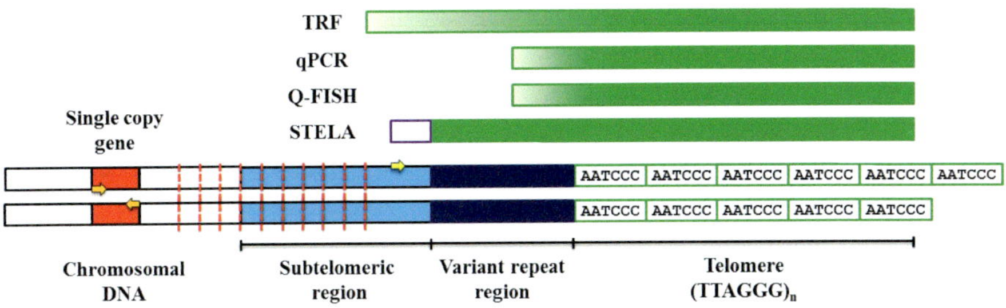

Fig. 1. Schematic representation of telomeric and subtelomeric regions from a chromosome showing the telomere length estimation of TRF, qPCR, Q-FISH and STELA methods. Telomeres are represented as green empty boxes with the TTAGGG and AATCCC sequence repeat, variant region (composed of degenerate TTAGGG repeats and telomere associated repeats) is portray as dark blue boxes, the subtelomeric region is presented as light blue boxes, the rest of chromosomal DNA is reproduced as a white box where a single copy gene (red box) is found and can be amplified by PCR using specific primer (orange arrows). Frequent cutting restriction endonucleases are depicted as red dotted lines. The telomere length estimation of the four techniques is represented as the green coloured boxes on the top: TRF includes part of the variant repeat region and subtelomeric region which are not digested by the frequent cutting restriction endonucleases. PCR based methods compare the amplification of telomeric DNA with the amplification of a single copy gene (red box) and can include part of the variant repeat region. Q-FISH probes recognize pure telomeric repeats but tandems of these can be also part of the variant repeat region. STELA uses a specific primer of the subtelomeric region (yellow arrow), but since this region is well characterized (dark blue empty box) it can be subtracted to the telomere length quantification. Figure modified from Aubert et al. (Aubert et al. 2008).

in senescence and aging related diseases (Vera and Blasco 2012, Aubert et al. 2012), not being able to detect them could miss critical information.

In order to solve some of the limitations listed above some modifications of TRF analysis were developed, for example performing telomere length determination by TRF of specific chromosomes after being sorted by flow cytometry (Vera et al. 2012), the use of pulsed-field electrophoresis to increase gel resolution (Lauzon et al. 2000) or a slot blot TRF analysis where the amount of genomic DNA could be reduced to 15 ng (Bryant et al. 1997, 1999).

PCR-based methods

As already mentioned previously, quantification of telomere length by TRF is a lengthy protocol that requires large amounts of DNA. To overcome these limitations, several PCR-based methods to measured telomere length have been developed.

Given that telomere consists of a repeating TTAGGG sequence, the primers required for telomere amplification are complementary thus favouring the formation of primer dimers rather than the amplification of the target DNA. This problem was overcome by the designing of primers that bind to the C- and G-rich segments, but are mismatched at the other bases, thus allowing to optimize a high-throughput monoplex real-time quantitative PCR (qPCR) (Cawthon 2002). For this approach it is necessary:

1. To obtain the threshold cycle (C_T) values corresponding to telomere fluorescent signal (T) of DNA samples in a PCR plate.
2. To obtain the fluorescent signal (C_T values) of a single-copy nuclear gene (S) of the same DNA samples, in another PCR plate.
3. To calculate the T/S ratio as an index of the mean telomere length for each DNA sample.

The precision of the assay is compromised, among others, by the necessity of performing T and S amplifications in different plates, thus favouring variations in pipetting of DNA.

Later, the original method was improved by developing a monochrome multiplex qPCR (Table 1) (Cawthon 2009). In this updated approach, both the telomeric ends and the single-copy nuclear gene (for example albumin gene) are amplified within the same tube. This strategy amplifies two amplicons with relevant differences that may be separately identified and quantified in different steps within the same tube at the same experiment (Fig. 2):

1. The telomeric amplicon, which possesses a final size of 79 bp (Fig. 3) and a melting temperature relatively low (80–82ºC), is quantified at earlier cycles of the reaction, just when the single-copy nuclear amplicon levels are not yet detectable (Fig. 2).
2. The single-copy amplicon, which possesses a size of 98 bp and a melting temperature (90–92ºC) higher than the telomeric one, is quantified at later cycles and at higher temperature than that used for quantification of telomeric amplicon (Fig. 2). This increase in the quantification temperature allows measuring only the single-copy amplicon-generated fluorescent signal, avoiding the interference of the much more abundant telomeric amplicon-signal.

This improved protocol reduces the operator-generated variability (Vera et al. 2012), and allows that the ratios between telomere and single-copy gene measurements are independent from small variations in pipetting volume. Besides is a cheaper method which exhibits a better correlation with other methodological approaches (Cawthon 2009).

The strength of PCR methods is that they are able to quantify the telomere length in small amount of DNA (ng levels or lower) and may be used to perform

Table 1. Primer sequences in monochromatic multiplex real-time quantitative PCR.

Primer	**Nucleotide sequence**
AlbuF	5'-**CGGCGGCGGGCGGCGCGGGCTGGGCGG**AAATGCTGCACAGAATCCTTG-3'
AlbdR	5'-**GCCCGGCCCGCCGCGCCCGTCCCGCCG**GAAAAGCATGGTCGCCTGTT-3'
TelgF	5'-ACACTAAGGTTTGGGTTTGGGTTTGGGTTTGGGTTAGTGT-3'
TelcR	5'-TGTTAGGTATCCCTATCCCTATCCCTATCCCTATCCCTAACA-3'

Bold nucleotides represent noncomplementary GC clamps, and allow to relevantly increase the melting temperature of the albumin amplicon.

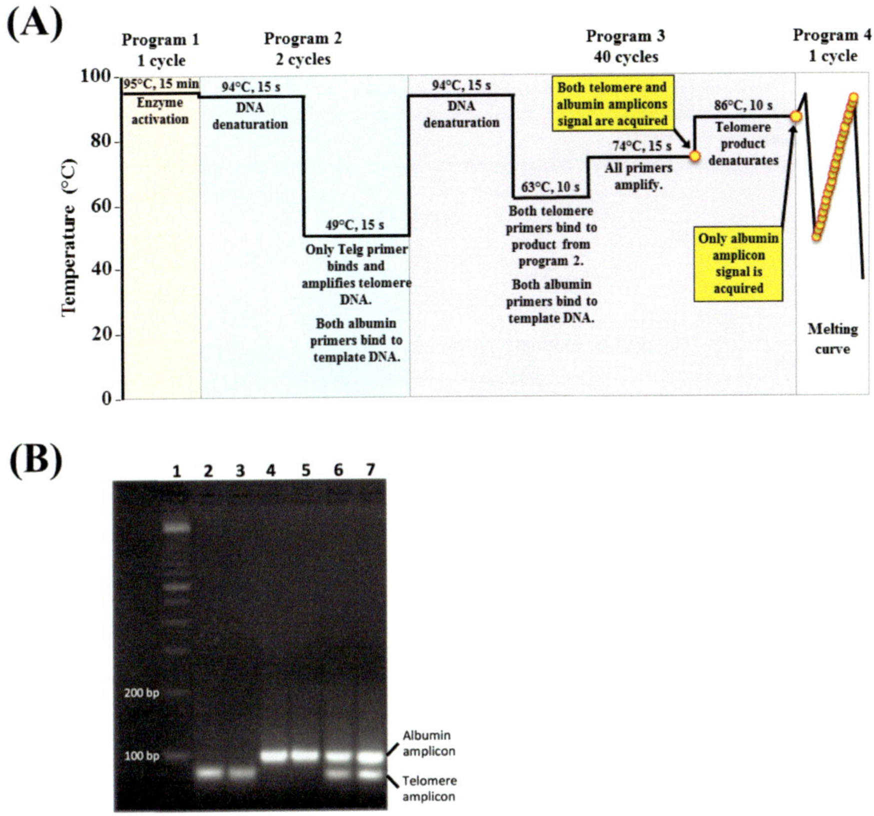

Fig. 2. (A) Schematic of the monochromatic multiplex real-time quantitative PCR cycling profile, describing each stage. The yellow circles represent fluorescence signal acquisitions. Figure modified from Hsieh et al. (Hsieh et al. 2016). (B) PCR products on 4% agarose gels. Lane 1 was loaded with a 100 bp-ladder. Lanes 2 through 7 are PCR products after 40-cycle profile. Lanes 2 and 3 show 79 bp-amplicon generated only with telomere primers. Lanes 4 and 5 show 98 bp-amplicon generated only with albumin primers. Lanes 6 and 7 show both amplicons generated with all primers.

quantifications in DNA biobanks. Also these methodologies are fast, relatively low cost, do not require the use of costly or specialized equipment and are easily amenably for high-throughput performance, which have make them widely used in genetic and epidemiological studies. Besides, PCR approaches also allow solving the variations associated with subtelomeric regions, as this region is not included in telomere length quantification (Fig. 1).

However, PCR methods do not provide an absolute telomere length calculation in kilobases, although this problem was solved by the use of a standard curve of known telomere length (Montpetit et al. 2014). Another limitation of PCR is that it only provides mean telomere length estimation of a sample (like TRF), not being able to measure telomere length of a single cell or being chromosome specific, furthermore it is not able to detect the presence of short telomeres (Vera and Blasco 2012).

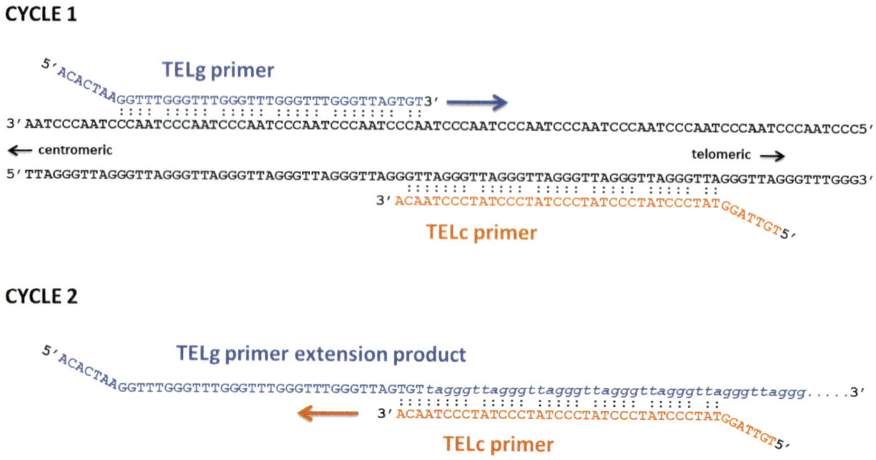

Fig. 3. Strategy for the generation of telomere amplicon. In cycle 1 only TELg primer (blue) is able to generate an extension product from original DNA template. TELc (orange) is not able to extend any product. In cycle 2 TELc primer may use TELg primer extension product as template. After this point, both telomere primers produce new extension products in each cycle to finally generate the 79 bp-telomere amplicon.

Finally, the greatest shortcoming of PCR approaches is the relatively high variability between different experiments performed by different laboratories, which limits the results comparisons between distinct studies (Montpetit et al. 2014, Vera et al. 2012). Classically, the intra- and inter-assay variation coefficients of the qPCR are higher as compared with other techniques. In a preliminary fully blinded study, performed by two laboratories, the coefficient of variation was higher in the qPCR (6.45%) than in the TRF (1.45%) (Aviv et al. 2011). However, a more recent blinded study conducted by 10 laboratories reported no significant differences in intra- and inter-assay variation coefficients among qPCR, Southern blotting or Single Telomere Length Analysis (STELA) (Martin-Ruiz et al. 2015). In this latter study, both Cawthon's qPCR approaches appeared to exhibit a similar reproducibility (Hsieh 2016). A better understanding of sources of variability—reagents and PCR machines—will allow us to design optimized protocols able to lower qPCR variability to the range reported for other techniques.

Quantitative fluorescence in situ hybridization

To overcome some of the drawbacks of TRF assay a new method for telomere length measurement was developed: the Quantitative Fluorescence *in situ* Hybridization (Q-FISH) technology (Lasdorp et al. 1996, Rufer et al. 1998). This technique is based on the hybridization of a fluorescent $(CCCTAA)_3$ probe complementary to the TTAGGG repeats sequence of telomeric DNA, due to each labelled probe binds to three TTAGGG telomeric repeats there is a direct proportion between fluorescence intensity and telomere length, which can be detect and quantify by digital fluorescence microscopy (Metaphase Q-FISH, Interphase Q-FISH and its improvements) or flow cytometry (Flow-FISH). The strict hybridization conditions of Q-FISH only allows the fluorescent probe to unite to perfect TTAGGG telomeric repeats, avoiding the inclusion of the subtelomeric region in the quantification of telomere length, although tandems of perfect TTAGGG repeats can be also part of the variant repeat region and might lead to a slight overestimation of telomere length (Fig. 1).

Instead of using DNA as a substrate like TRF, STELA and PCR techniques, Q-FISH assays are performed directly in cells which could be fresh, frozen, fixed or embedded in paraffin (Montpetit et al. 2014). Thanks to this Q-FISH techniques are able to determine telomere length of a single cell (Flow-FISH, Interphase Q-FISH and its improvements) or even the telomere end of a single chromosome (Metaphase Q-FISH) (Vera and Blasco 2012).

The most frequently fluorescent probe used is a synthetic peptide nucleic acid (PNA) molecule instead of a fluorescent oligonucleotide probe. In the PNA molecule the negatively charged phosphate ribose/deoxyribose backbone is replaced by a neutral N-2 amine ethyl-glycine structure, increasing the probe stability and making the binding to the telomeric DNA sequence much more efficient and specific (Vera and Blasco 2012, Lauzon et al. 2000). Furthermore PNA probes are able to hybridize under low ionic strength conditions, in which DNA-DNA and RNA-DNA interactions are not favoured, avoiding reannealing of the telomeric DNA strands and leading to stronger fluorescent signal and more reproducible results (Poon and Landsdorp 2001).

Metaphase Q-FISH

Metaphase Q-FISH was the first of the Q-FISH techniques developed in 1996 (Lasdorp et al. 1996). Briefly, in this protocol the sample is prepared by arresting the cell cycle in metaphase and spreading the cells out on a slide, so each chromosome could be identified and karyotyped. Then the sample is fixed and DNA is denatured, once the double strand is open the PNA probe (usually labelled with fluorescein isothiocyante [FITC], 5-Carboxytetramethylrhodamine [TAMRA] or Cyanine 3 [Cy3]) is hybridized to the telomeric DNA target and the remaining chromatin is counterstained with nonspecific DNA fluorescent dye (like 4',6-diamidino-2-phenylindole [DAPI] or propidium idodide [PI]). Finally images are visualized and acquired with a fluorescence microscopy or with a digital imaging system to quantify the fluorescent signals from the telomeres and the DNA. Using control samples of known length the fluorescent intensity signal can be converted in to kilobases values (Montpetit et al. 2014, Baerlocher and Lansdorp 2004, Lin and Yan 2005). When analysing the data it is important to achieve appropriate telomere segmentation, that is, to correctly delimit the area of the PNA probe spot from the background DNA, which will be used to estimate the length of each individual telomere (Poon and Landsdorp 2001). The final result is usually a histogram with the distribution of telomere length frequencies (Vera and Blasco 2012).

The main advantage of the metaphase Q-FISH is its capacity to determine the size of each single telomere from a specific chromosome, with highly accurate and sensitive estimation of telomere length (its detection limit has been determined in less than 0.15 kb) (Vera and Blasco 2012). Due to the use of metaphase arrested cells and its high sensitivity, this technique is able to detect small telomeres and also telomere free ends (chromosomes with critically short ends or uncapped chromosomes) (Montpetit et al. 2014), as it has been said before the relevance of short telomeres has been proved in senescence and aging related diseases, so metaphase Q-FISH provides critical information in this respect. Furthermore it combines the telomere length quantification with karyotyping analysis and the study of chromosomal abnormalities like chromosome fusion events (Vera et al. 2012, Samassekou et al. 2010). This technique has been optimized for many settings and is able to measure telomere length in rare cells using a limited number of cells (Poon and Landsdorp 2001), to date metaphase Q-FISH is the finest and most accurate technology available to quantify individual telomeres per cell.

However metaphase Q-FISH has several disadvantages that must be taken into account, the requirement of cells arrested at metaphase makes impossible to measure telomere length in cells that are not mitotically active, like senescent cells, and very difficult in samples with a low proliferation rate (Montpetit et al. 2014). This method is very time consuming, costly, labor intensive (it demands technical expertise in karyotyping, knowledge of chromosomal banding patterns and the preparation of the metaphase spreads samples), which makes it not suitable for large epidemiological studies. Also, it requires appropriate fluorescence microscope and digital camera equipment, which is expensive and needs a properly trained technician (Baerlocher and Lansdorp 2004) and due to the procedure of sample preparation there is a limitation in the fluorophores that can be used to label the telomeric probe (Vera and Blasco 2012).

Some modifications from the original metaphase Q-FISH method have been developed (Lin and Yan 2005):

- TELI-FISH (TEL-FISH and Telomere/Immunostaining-FISH): This method combines measurement of telomere length Q-FISH with immunostaining formalin-fixed, paraffin-embedded samples (Meeker et al. 2002).
- T/C-FISH (telomere/centromere-FISH): This technique uses the centromere of chromosome 2 as an internal reference and is able to measure very precisely individual telomeres at every single chromosome arm (Perner et al. 2003).

Interphase Q-FISH

Interphase Q-FISH is based on the same principles of the metaphase Q-FISH method, but using interphase cells instead of metaphase chromosomes. However, like in metaphase Q-FISH, PNA probes can be used for interphase Q-FISH and there is a wide range of samples types that can be assessed (blood cells, fresh, frozen, fixed or embedded in paraffin cells) (Montpetit et al. 2014).

Usually the interphase Q-FISH procedure involves a double staining with two fluorophores; one is the labelled PNA probe that binds the telomeric DNA sequence and the second is a centromeric probe, so both target signals can be quantified and the telomere fluorescent signal normalized with the centromeric probe (Vera et al. 2012).

As interphase Q-FISH does not require the use of metaphase-arrested cells, it solves the shortcoming of metaphase Q-FISH concerning its limitation to mitotically active cells and is less labor intensive and time consuming. But it has the drawback of neither being able to measure the size of single telomeres of individual chromosomes nor to detect telomere free ends. Instead interphase Q-FISH provides mean length estimation of aggregations of few telomeres (defined as telomere spots), which gives a good estimation of the frequency of short telomeres (Vera and Blasco 2012).

Interphase Q-FISH method has been improved in several ways:

- Fiber FISH: This high-resolution method includes different protocols where the molecular probe binds to released chromatin or DNA fiber. Fiber FISH is specially used for physical mapping and genome/chromosome structure studies (Heng et al. 1998).
- Tissue Q-FISH cytometry: In this semiautomated variant of interphase Q-FISH telomere length is measured in tissue sections using a telomere/centromere fluorescent signal ratio to normalize the signal (Aida et al. 2007, Aida et al. 2008).
- HT Q-FISH (High-throughput Q-FISH): An improvement of interphase Q-FISH (Canela et al. 2007), HT Q-FISH is based on metaphase Q-FISH method but assessing interphase nuclei with high-throughput microscopy in a 96 well format. This technology has been validated and shows a good correlation with TRF, conventional metaphase Q-FISH and flow fish, being as accurate and sensitive as metaphase Q-FISH. The major advantages of HT Q-FISH is that is a precise and sensitive technique able to measure telomere length in almost any cell type, it is automated and fast (can process 96 samples with 1000 nuclei per sample in 2 hours) which makes it suitable for large epidemiological studies; capable of calculating the frequency of short telomeres and even the study of individual telomere spots in a cell. However it still has some limitations, it cannot detect telomere free ends and requires confocal microscopy equipment (Vera and Blasco 2012).

Flow-FISH

Flow FISH methodology combines the use of cell sorting by flow cytometry with the basis of FISH applied to telomere length measurement (hybridization of a fluorescent labelled probe, usually a FITC-PNA molecule, which binds to the telomeric repeats sequence) (Rufer et al. 1998). However, unlike interphase Q-FISH and metaphase Q-FISH, it can only assess cells in suspension.

Telomere length measurement by flow FISH protocol is divided in six steps (Lauzon et al. 2000, Baerlocher et al. 2002):

1. Cell separation and preparation: Isolation of the nucleated cells from the sample (blood or tissue) and its preparation for flow cytometry. As Flow-FISH requires cells in suspension, this technique has been widely used for the quantification of telomere length in hematopoietic cell subtypes. When assessing whole blood or bone marrow, white blood cells must be purified first, since hemoglobin from red blood cells could interfere quenching the fluorescent signal. To achieve this, erythrocytes have to be lysed (usually with chemical reagents like NH_4Cl) without destroying white blood cells.

2. DNA denaturation: In this step the double strand DNA molecule is open to allow the binding of the probe. The denaturation process involves the application of heat and formamide during a certain time without destroying the DNA or the cell. Since this process of denaturation could also affect the proteins, there is a limitation in the type of fluorophores that can be link with the telomeric probe, it has to be a non-proteic molecule.
3. Hybridization with fluorescent probe: In this step the fluorescent probe gets inside the cell to bind to the telomeric DNA. Sometimes fixation reagents could reduce cell permeability and diminish probe access, so it must be reach a balance between enough fixation to maintain cell structure and probe penetration. Like in interphase Q-FISH and metaphase Q-FISH, PNA probes are the best choice for flow FISH due to its specific binding under low ionic strength conditions, which also avoid DNA reannealing. In fact PNA probes penetration inside the cell is increased after the heat and formamide treatment, since it enhances membrane permeabilization.
4. Washes: In this step unbound PNA probe is eliminated to reduce background signal, if it is not removed properly the telomeric probe can bind unspecifically to cellular organelles and membranes, leading to an overestimation of telomere length. To solve this, it has been proposed the use of nuclei instead of whole cells for the measurement of telomere length by flow FISH (Baerlocher et al. 2006).
5. DNA counterstaining: Using nonspecific DNA fluorescent dye (for example DAPI or PI), whose signal is used to normalize telomere probe quantification.
6. Acquisition and analysis on flow cytometer: Cells are sorted and pass one by one through the lasers and fluorescent signal from the PNA probe and the DNA dye is measured. Measuring control cells of known telomere length allows calculating absolute telomere length values from the fluorescent signal of the samples. Depending of the stage of the cell cycle the number of telomeres per nuclei vary (because DNA is duplicated in S phase), affecting the amount of fluorescent signal, to solve this problem and homogenize the analysis it has been suggested to quantify telomere length only in cells during G_0 and G_1 phase (Hultdin et al. 1998).

Every step and all experimental conditions (temperature, reagent concentrations, incubation times…) have to be optimized for accuracy and reproducibility of flow FISH. Target cell destruction, incomplete DNA denaturation and unspecific binding of the telomeric probe must be avoided.

The greatest strength of Flow FISH is its capacity to measure telomere length in single cells after having sorted the cells into different subpopulations according to size, granulation or antibody detection, even in the same sample. This technique requires a relatively low number of cells (approximately 100000) per sample (Baerlocher et al. 2002), some of the procedure steps can be performed automatically and is less time consuming comparing with metaphase Q-FISH. These characteristics make flow FISH an accurate method suitable for medium scale studies.

However flow FISH has some limitations; first, it is capable of measuring mean values of telomere length per cell only and cannot detect individual telomeres or telomere free ends. Second, sample cells need to be fixed, if not they will form clumps or be destroy, but the method is sensitive to the use of fixation chemicals, thus these can produce the loss of cell surface epitopes or affect the hybridization efficiency. Finally, flow FISH is quite labor intensive, it requires high skill level to be correctly performed and flow sorting equipment.

Some modifications of the original protocol have been done to improve flow FISH:

- Automated multicolor flow FISH: In this technique flow FISH has been adapted to a 96 well format partially automated for higher throughput analysis, with the use of an internal control in each sample and combining the procedure with limited immunophenotyping (antibody staining) (Baerlocher et al. 2006). Actually automated multicolor flow FISH is the most sensitive and fastest method for telomere length measurement in white blood cells, reducing the number of cells required to one thousand and having a resolution of 0.2–0.5 kb (Baerlocher et al. 2003).
- Three dimension flow FISH: With this method telomere length can be quantified and its spatial organization studied through the cell cycle, by defining a distribution parameter (ρ_T) that defines the shape (spherical or disk-like) distribution of the telomere (Vermolen et al. 2005).

Single Telomere Length Analysis

STELA analysis was developed to solve the limitations of TRF and Q-FISH and initially use to study the short arms of sex chromosomes (Baird et al. 2003).

STELA is a ligation mediated PCR method whose procedure is divided in five steps (Vera et al. 2012, Lin and Yan 2005, Baird et al. 2003):

1. Annealing of "telorette" linker: Oligonucleotide molecules compose of 7 nucleotides complementary to the TTAGGG telomeric repeats follow by a 20 bases non-complementary tail to the G rich 3' overhang.
2. Ligation of the "telorette" linker: The "telorette" oligonucleotide is ligated to the C rich 5' end strand of the chromosome, tagging the telomere with the non-complementary tail.
3. PCR amplification: Using a primer ("teltail") complementary to the "telorette" non-complementary tail and a chromosome-specific primer for the subtelomeric region, an amplicon can be obtained by PCR whose length is approximately equal to the telomere target. This technique can be used also to study polymorphisms in the subtelomeric region using an allele-specific subtelomeric primer.
4. DNA separation by agarose gel electrophoresis.
5. Detection by Southern blot: The developed bands represent the telomere length of the specific telomere target.

The biggest advantage of STELA technique is its capacity to measure the length of telomeres in specific chromosomes and detecting critically short telomeres even with small amounts of starting material (few picograms of DNA or as few as 50 cells), which makes this method a good choice for the study of rare cells and non-dividing senescent cells (Vera et al. 2012). STELA is a highly accurate technique that is able to measure telomere lengths ranging from 406 bp to 20 kb and has a good correlation with TRF. Due to the PCR amplification and the specific subtelomeric primer STELA is highly specific, can be used to study allelic differences in individual telomeres and unlike TRF, STELA is not so affected by the inclusion of subtelomeric length in telomere length quantification (since the subtelomeric region included in the amplification is well characterized its contribution can be subtracted) (Lin and Yan 2005)(Fig. 1). Finally it requires no specialized equipment for telomere analysis.

Due to the requirement that the sequence target for the subtelomeric primer has to be chromosome-specific STELA method can only be used for a small group of chromosomes (XpYp, 2p, 11q, 12q, and 17p), making this the great limitation of this technology (Montpetit et al. 2014). STELA has an upper limit of 20 kb in the measurement of telomere length, making this method not suitable for mice samples (Vera et al. 2012) and is sensitive to the starting amount of DNA in the sample (an excess of template could lead to the obtaining of smears due to unfinished amplicons). It is a labor intensive and technically challenging technique, that requires good optimization of the procedure so it has a low-throughput nature and it is not applicable to large studies (Vera and Blasco 2012).

Universal STELA

Regarding the relevance of short telomeres in age related diseases and as only a few chromosomes can be asses with STELA, it is clear that this method is limited to study the whole state of short telomeres of a cell. To overcome this shortcoming an improvement of conventional STELA was developed: the universal STELA (Bendix et al. 2010).

Briefly universal STELA protocol has 6 steps:

1. Genomic DNA digestion: Using two restriction enzymes to cut sample DNA into pieces of 100 to 3000 bp leaving sticky overhangs. For the intragenic fragments both ends are left with the same sticky overhang while the telomere fragments has only one since the 3’ single strand end cannot be cut. MseI and NdeI are the best combination of enzymes since both minimize the size of the subtelomeric region leading to a minor overestimation of telomere length.
2. Annealing and ligation of the panhandle linker and the “telorette” linker: The panhandle linker is a double-stranded oligo that binds to the sticky overhangs produced by the restriction enzymes and the “telorette” linker is the same as conventional STELA and binds to the G rich 3’ overhang of the telomeres. The oligos are ligated to the DNA fragments leaving the intragenic fragments with panhandle linkers in both ends while the telomeric fragments have only panhandle linker and one “telorette” linker bind to the 3’ overhang. The panhandle linkers are designed to form a loop when two oligos are present at both ends of a DNA fragment, not allowing the intragenic fragments to be amplified by PCR reaction.
3. Fill-in step: The sticky overhangs of the panhandle linkers are completed and become double stranded, so it can serve as a template for PCR in the telomeric fragment and allows the formation of the loop in the intragenic fragments during the PCR.

4. PCR amplification: This step is like the original STELA, but in this case not one but all the telomeres of each chromosome are amplified, while the intragenic fragments are being blocked by the loop.
5. DNA separation by agarose gel electrophoresis.
6. Detection by Southern blot: The developed bands represent the telomere length of each specific telomere target.

Universal STELA allows the detection of critical short telomeres in all chromosomes using small amounts of DNA, solving the limitation of conventional STELA. However is less useful when measuring mean telomere length and like conventional STELA is not able to detect telomeres longer than 20 kb (indeed it has been reported that few telomeres longer than 8 kb can be detected with universal STELA) (Montpetit et al. 2014).

Other Methods

Hybridization protection assay (HPA)

This methodology was first used by Nakamura et al. (Nakamura et al. 1999) to measure telomere length in 1999, originally it was used to quantify telomerase activity. In HPA sample DNA is hybridized with an acridinium ester labelled probe for telomere target and with another for *Alu* sequences. Both chemiluminescence signals are quantified by a luminometer and a ratio of telomere to *Alu* sequences is calculated.

HPA method is simple, easy to handle, relatively quick (it requires approximately 45 min), sensitive, quantitative and reproducible. It does not include the subtelomeric region in telomere length quantification. HPA does not require high quality DNA (of high purity and unsheared) nor large amounts of starting DNA (the lower limit is 1000 cells) (Lin and Yan 2005) and can be used to assess clinical samples like biopsied or resected human soft tissues, cell lysates, body fluid or samples obtained by endoscopy.

However HPA only is able to estimate the mean telomere length of a cell population and cannot distinguish specific chromosomes, individual telomeres or cells. Sometimes is difficult to interpret the ratio results due to variation in the *Alu* sequences between samples and converting it into an absolute kilobase measurement. Because of these limitations HPA is not frequently used in telomere length quantification (Montpetit et al. 2014).

There is an alternative version of HPA called hybridization assay, in which the oligonucleotide probes are labelled with acetylcholinesterase and the enzymatic reaction is quantified using a colorimeter (Vera and Blasco 2012).

Primed in situ (PRINS)

PRINS is a complementary technology for Q-FISH methods instead of using PNA probes. Oligonucleotide $(CCCTAA)_7$ primers are annealed to telomeres of spread chromosomes (from metaphase or interphase cells), amplified by DNA polymerase using fluorochrome labelled nucleotides and after elongation the fluorescent signal is measured using fluorescence microscopy (Montpetit et al. 2014).

PRINS allows measurements of individual telomeres and is able to detect critically short telomeres. Some modifications of the original method have been developed to overcome its limitations, like multiple cycles amplification to enhance fluorescent signal, enzymatic detection, the use of dideoxynucleotides for a more specific labelling and the double-strand PRINS method (which uses two primers to elongate both telomeric DNA strands) to increase fluorescent signal (Lin and Yan 2005).

Conclusion

As it has been described throughout the chapter, there are several methods available for telomere length measurement, each one of them with its strong points and limitations and there is not a unique technique that stands out being at the same time accurate, sensitive, fast and easy. So choosing the appropriate method is fundamental when performing the experiment and some points have to be taken into account.

One important fact when choosing the right technique is the type of measurement obtained, some methods are only able to quantified mean telomere length of a sample (like TRF and qPCR), of single cells (flow FISH) or of telomere spots (interphase Q-FISH). STELA can measure the length of single telomeres of certain chromosomes, but only metaphase Q-FISH is able to measure the length of single telomeres of all chromosomes. In the recent years it has been highlighted the relevance of short telomeres, so the ability of detect extremely short telomeres or free ends (metaphase Q-FISH, STELA and Universal STELA) is a critical point when selecting the right method.

According to the type of sample used all Q-FISH methods requires living cells while TRF, qPCR and STELA use DNA directly. It is important to remember that metaphase Q-FISH requires mitotically active cells and cannot be performed with senescence cells. Also when analysing mice samples, telomere quantification is not suggested to be performed with STELA and Universal STELA due to their upper limit detection of 20 kb.

The availability of starting material is another important factor, TRF is the method which requires the greatest amount of sample while STELA, metaphase Q-FISH and interphase Q-FISH require the lowest. Also the amount of samples to be studied should be taken into account, TRF, metaphase Q-FISH and STELA can only process a low number of samples while qPCR, interphase Q-FISH and flow FISH are suitable for large scale studies.

Acknowledgments

Álvaro Pejenaute acknowledges support from the FPU grant (Formación de Profesorado Universitario FPU12/01994) from the Spanish Education, Culture and Sport Ministry (Ministerio de Educación, Cultura y Deporte).

This work was supported by the Spanish Ministry of Economy and Competitiveness (SAF2016-79151-R).

Keywords: Polymerase chain reaction, quantitative fluorescence *in situ* hybridization, terminal restriction fragment analysis, STELA

References

Aida, J., N. Izumiyama-Shimomura, K. Nakamura, A. Ishii, N. Ishikawa, N. Honma et al. 2007. Telomere length variations in 6 mucosal cell types of gastric tissue observed using a novel quantitative fluorescence *in situ* hybridization method. Hum. Pathol. 38: 1192–1200.

Aida, J., N. Izumiyama-Shimomura, K. Nakamura, N. Ishikawa, S.S. Poon, M. Kammori et al. 2008. Basal cells have longest telomeres measured by tissue Q-FISH method in lingual epithelium. Exp. Gerontol. 43: 833–839.

Aubert, G., M. Hills and P.M. Lansdorp. 2012. Telomere length measurement-caveats and a critical assessment of the available technologies and tools. Mutat. Res. 730: 59–67.

Aviv, A., S.C. Hunt, J. Lin, X. Cao, M. Kimura and E. Blackburn. 2011. Impartial comparative analysis of measurement of leukocyte telomere length/DNA content by Southern blots and qPCR. Nucleic Acids Res. 39: e134.

Baerlocher, G.M., J. Mak, T. Tien and P.M. Lansdorp. 2002. Telomere length measurement by fluorescence *in situ* hybridization and flow cytometry tips and pitfalls. Cytometry. 47: 89–99.

Baerlocher, G.M. and P.M. Lansdorp. 2003. Telomere length measurements in leukocyte subsets by automated multicolor Flow-FISH. Cytometry. 55A: 1–6.

Baerlocher, G.M. and P.M. Lansdorp. 2004. Telomere length measurements using fluorescence *in situ* hybridization and flow cytometry. Methods Cell Biol. 75: 719–750.

Baerlocher, G.M., I. Vulto, G. de Jong and P.M. Lansdorp. 2006. Flow cytometry and FISH to measure the average length of telomeres (flow FISH). Nat. Protoc. 1: 2365–2376.

Baird, D.M., J. Rowson, D. Wynford-Thomas and D. Kipling. 2003. Extensive allelic variation and ultrashort telomeres in senescent human cells. Nat. Genet. 33: 203–207.

Bendix, L., P.B. Horn, U.B. Jensen, I. Rubelj and S. Kolvraa. 2010. The load of short telomeres, estimated by a new method, Universal STELA, correlates with number of senescent cells. Aging Cell. 9: 383–397.

Bryant, J.E., K.G. Hutchings, R.K. Moyzis and J.K. Griffith. 1997. Measurement of telomeric DNA content in human tissues. Biotechniques. 23: 476–478.

Bryant, J.E. 1999. Sensitive Method for Measurement of Telomeric DNA Content in Human Tissues. US Patent.

Canela, A., E. Vera, P. Klatt and M.A. Blasco. 2007. High-throughput telomere length quantification by FISH and its application to human population studies. Proc. Natl. Acad. Sci. USA. 104: 5300–5305.

Cawthon, R.M. 2002. Telomere measurement by quantitative PCR. Nucleic Acids Res. 30: e47.

Cawthon, R.M. 2009. Telomere length measurement by a novel monochrome multiplex quantitative PCR method. Nucleic Acids Res. 37: 1e7.

Heng, H.H. and L.C. Tsui. 1998. High resolution free chromatin/DNA fiber fluorescent *in situ* hybridization. J. Chromatogr. A. 806: 219–229.

Kimura, M., R.C. Stone, S.C. Hunt, J. Skurnick, X. Lu, X. Cao et al. 2010. Measurement of telomere length by the Southern blot analysis of terminal restriction fragment lengths. Nat. Protoc. 5: 1596–1607.

Hsieh, A.Y., S. Saberi, A. Ajaykumar, K. Hukezalie, I. Gadawski, B. Sattha et al. 2016. Optimization of a relative telomere length assay by monochromatic multiplex real-time quantitative PCR on the Light Cycler 480. Sources of variability and quality control considerations. J. Mol. Diagn. 18: 425–437.

Hultdin, M., E. Grönlund, K. Norrback, E. Eriksson-Lindström, G. Roos and T. Just. 1998. Telomere analysis by fluorescence *in situ* hybridization and flow cytometry. Nucleic Acids Res. 26: 3651–3656.

Lauzon, W., J. Sanchez Dardon, D.W. Cameron and A.D. Badley. 2000. Flow cytometric measurement of telomere length. Cytometry. 42: 159–164.

Lansdorp, P.M., N.P. Verwoerd, F.M. van de Rijke, V. Dragowska, M.T. Little, R.W. Dirks et al. 1996. Heterogeneity in telomere length of human chromosomes. Hum. Mol. Genet. 5: 685–691.

Lin, K.W. and J. Yan. 2005. The telomere length dynamic and methods of its assessment. J. Cell. Mol. Med. 9: 977–989.

Martin-Ruiz, C.M., D. Baird, L. Roger, P. Boukamp, D. Krunic, R. Cawthon et al. 2015. Reproducibility of telomere length assessment-An international collaborative study. Int. J. Epidemiol. 44: 1749–1754.

Meeker, A.K., W.R. Gage, J.L. Hicks, I. Simon, J.R. Coffman, E.A. Platz et al. 2002. Telomere length assessment in human archival tissues: combined telomere fluorescence *in situ* hybridization and immunostaining. Am. J. Pathol. 160: 1259–1268.

Montpetit, A.J., A.A. Alhareeri, M. Montpetit, A.R. Starkweather, L.W. Elmore, K. Filler et al. 2014. Telomere length: a review of methods for measurement. Nurs. Res. 63: 289–299.

Nakamura, Y., M. Hirose, H. Matsuo, N. Tsuyama, K. Kamisanco and T. Ide. 1999. Simple, rapid, quantitative, and sensitive detection of telomere repeats in cell lysate by a hybridization protection assay. Clin. Chem. 45: 1718–1724.

Perner, S., S. Brüderlein, C. Hasel, I. Waibel, A. Holdenried, N. Ciloglu et al. 2003. Quantifying telomere lengths of human individual chromosome arms by centromere-calibrated fluorescence *in situ* hybridization and digital imaging. Am. J. Pathol. 163: 1751–1756.

Poon, S.S. and P.M. Lansdorp. 2001. Measurements of telomere length on individual chromosomes by image cytometry. Methods Cell Biol. 64: 69–96.

Rufer, N., W. Dragowska, G. Thornbury, E. Roosnek and P.M. Lansdorp. 1998. Telomere length dynamics in human lymphocyte subpopulations measured by flow cytometry. Nat. Biotechnol. 16: 743–747.

Samassekou, O., M. Gadji, R. Drouin and J. Yan. 2010. Sizing the ends: normal length of human telomeres. Ann. Anat. 192: 284–291.

Vera, E. and M.A. Blasco. 2012. Beyond average: potential for measurement of short telomeres. Aging (Albany NY). 4: 379–392.

Vera, E., B. Bernardes de Jesus, M. Foronda, J.M. Flores and M.A. Blasco. 2012. The rate of increase of short telomeres predicts longevity in mammals. Cell Rep. 2: 732–737.

Vermolen, B.J., Y. Garini, S. Mai, V. Mougey, T. Fest, T.C.Y. Chuang et al. 2005. Characterizing the three-dimensional organization of telomeres. Cytometry. 67A: 144–150.

Chapter 4

Telomere Integrity and Length and Cancer Risk

Jian Gu and *Xifeng Wu**

INTRODUCTION

Telomeres consist of tandem arrays of short repetitive, non-coding sequences (TTAGGG in humans) and associated proteins at the ends of each chromosome (Blackburn 2005, Blackburn et al. 2006). Telomeres form a loop structure that caps the chromosomes and prevent chromosomes from erosion, end-to-end fusion, irregular recombination, and other damaging chromosomal events (Blackburn 2005, Blackburn et al. 2006). In somatic cells under normal physiological conditions, telomeres are progressively shortened during each mitotic cell division due to the "end replication problem", resulting in progressive shortening of the chromosomes after each cell division. Because telomeres consist of repetitive, non-coding sequences and sit at both ends of each chromosome, the cells can afford to lose a few hundred base pairs of telomeres without losing genetic information. However, when telomeres become too short, the loop structures and associated proteins are disrupted and the protective function of telomere is lost. Critically short telomeres can have dire consequences, causing apoptosis and replicative senescence, and have been linked to the development of many chronic diseases.

Telomere length is widely considered a marker of biological ageing. There have been increasing human population studies of telomere length, predominately measured in peripheral blood leukocytes due to the easy accessibility of samples. Although leukocyte telomere length (LTL) is generally inversely correlated with age, there is considerable inter-individual variation of LTL among people of the same ages (Aubert and Lansdorp 2008, Lansdorp et al. 1996). This inter-individual variation of LTL has been proposed to contribute to an individual's susceptibility to age-related

Department of Epidemiology, The University of Texas MD Anderson Cancer Center, Houston, TX; USA.
* Corresponding author: xwu@mdanderson.org

diseases. An individual's LTL at any given age is determined by a combination of genetic, environmental, and lifestyle factors (Lin et al. 2012). LTL is under strong genetic control with an estimated heritability of up to 80% from classic twin studies (Jeanclos et al. 2000, Slagboom et al. 1994). Recent genome-wide association studies (GWAS) have identified at least ten genetic loci associated with LTL (Codd et al. 2010, Codd et al. 2013, Levy et al. 2010, Mangino et al. 2015, Mangino et al. 2012, Pooley et al. 2013). On the other hand, a growing body of evidence has shown that LTL is modulated by a variety of environmental and lifestyle factors, including smoking, obesity, alcohol drinking, air pollution, sedentary lifestyle, and chronic stress, etc. (Lin et al. 2012, Ludlow and Roth 2011, Needham et al. 2013, Nettleton et al. 2008, Price et al. 2013, Sanders and Newman 2013, Valdes et al. 2005), many of which are common risk factors for cancer and other chronic diseases. It has been postulated that telomeres may serve as a biological mediator between these modifiable risk factors and diseases. Indeed, there have been numerous reports suggesting that short LTL is associated with increased risks of many chronic diseases, including cardiovascular diseases, cancer, diabetes, chronic obstructive pulmonary disease (COPD), and neurodegenerative diseases (Brouilette et al. 2007, Fyhrquist et al. 2013, Honig et al. 2012, Oeseburg et al. 2010, Sanders et al. 2012, Steffens et al. 2012, Weischer et al. 2013a, Wentzensen et al. 2011). In this chapter, we will focus on cancer risks and summarize our current knowledge of the genetics of LTL as well as the link of LTL with the risk of cancers.

Technical Issues and Study Design in Association Studies of LTL and Cancer Risk

The most commonly used methods for measuring LTL include Southern blot, quantitative PCR, and quantitative fluorescence *in situ* hybridization (qFISH) (Aubert et al. 2012). The Southern blot method is the gold standard for measuring LTL (Kim et al. 1994) with the lowest inter-assay variability (coefficient of variance typically $< 2\%$) among all the methods. However, the Southern blot method is time-consuming and labor-intensive and requires a large amount of input DNA (2–3 μg per sample) (Kimura et al. 2010), which limits its application in large scale population studies. Because of the low-throughput of Southern blot method and ever-increasing sample sizes in population studies, recent population studies of LTL have been mostly done with the high-throughput, quantitative real-time PCR method developed by Cawthon (Cawthon 2002). This method measures the relative average telomere length of each sample by determining the ratio of telomeric repeat copy number to single copy gene (e.g., human globulin) copy number (T/S ratio). Relative LTL produced from this PCR method has an excellent correlation with the absolute telomere length as determined by Southern blot method (Cawthon 2002). One of the major limitations of this method is its relatively high inter-assay variations and variability across different laboratories (Aviv et al. 2011), which may partially contribute to the inconsistent results regarding the association of LTL with cancer risks in the literature. Both the Southern blot method and the real time PCR method measure the mean LTL of all the chromosomes. In contrast, qFISH-based method utilizing common probes for telomere repeats can profile LTL in each individual chromosome and has been used widely in basic mechanistic research and a few small scale population studies (Aubert et al. 2012, Bruderlein et al. 2008, O'Sullivan et al. 2006, Zheng et al. 2011). But the limitations of low-throughput, requiring intact cells, and relatively high cost limit its application in large-scale population studies.

Study design is of paramount importance in association studies of LTL with cancer risks. The vast majority of published studies are cross-sectional studies measuring LTL at a single time point. Such studies require large sample sizes due to the substantial inter-individual variability of LTL and inter-assay variability of LTL measurement. Longitudinal studies measuring LTL at multiple time points within individuals can obtain LTL erosion rate and would only need five times smaller sample sizes than cross-sectional studies (Aviv et al. 2006). A more important issue of study design is that prospective studies are always preferred over retrospective study in disease association studies because of the inherited problem of reverse causality in retrospective case control studies. There is ample evidence that retrospective case control study may produce spurious associations or overestimate disease associations of LTL (Pooley et al. 2010, Weischer et al. 2013b, Wentzensen et al. 2011). In this chapter, whenever available, we will put more weight on large sample size and prospective studies.

Genetic variants associated with LTL

Telomere length is under strong genetic control. The relative length of each specific telomere is defined in the zygote, determined by inherited, telomere-near factors, and strictly maintained throughout life (Graakjaer et al. 2003, Graakjaer et al. 2006a, Graakjaer et al. 2006b). Martens et al. (Martens et al. 1998) showed that telomere length is similar in different tissues from the same individual and follows the same patterns in all individuals. Classic twin studies and linkage studies have estimated a heritability of up to 80% for LTL (Jeanclos et al. 2000, Slagboom et al. 1994, Vasa-Nicotera et al. 2005). In addition, Broer et al. (Broer et al. 2013) investigated the heritability and mode of inheritance of LTL in six large, independent cohort studies with a total of 19,713 participants. The meta-analysis estimate of LTL heritability was 0.70 (95% CI, 0.64–0.76). A stronger mother-offspring ($r = 0.42$; $P = 3.60 \times 10^{-61}$) than father-offspring correlation ($r = 0.33$; $P = 7.01 \times 10^{-5}$) was observed. Interestingly, a significant and quite substantial correlation in LTL between spouses ($r = 0.25$; $P = 2.82 \times 10^{-30}$) was seen, which appeared stronger in older spouse pairs than in younger pairs, suggesting shared environment also playing an important role in LTL dynamics. Because of the high genetic heritability of LTL, there have been many attempts to identify genetic loci that determine LTL. Putative loci were mapped to chromosome 3p26.1, 10q26.13, 12q12.22, and 14q23.2 in family-based linkage analyses (Andrew et al. 2006, Mangino et al. 2008, Vasa-Nicotera et al. 2005), but these regions need confirmation and candidate genes in these regions remain elusive. Several recent GWAS have identified single-nucleotide polymorphisms (SNPs) in at least ten regions that are associated with LTL (Codd et al. 2010, Codd et al. 2013, Levy et al. 2010, Mangino et al. 2015, Mangino et al. 2012, Pooley et al. 2013). Six of these loci contain candidate genes directly related to telomere homeostasis (*TERC, TERT, NAF1, OBFC1, CTC1* and *RTEL1*). Recent studies have found some of these SNPs were associated with the risks of different cancers (Bojesen et al. 2013, Codd et al. 2013, Du et al. 2015, Machiela et al. 2015, Pooley et al. 2013, Shi et al. 2013, Walsh et al. 2014), supporting a link of LTL with cancer risks.

Association of LTL with cancer risks

Since aging is the most important risk factors for many chronic diseases, short LTL, as a marker of biological aging, may predispose individuals to increased risks of aging-related chronic diseases, including cancer. Indeed, numerous association studies have shown significant associations of short LTL with the risks of a variety of chronic diseases, such as cardiovascular diseases, cancer, diabetes, COPD, and neurodegenerative (Brouilette et al. 2007, Fyhrquist et al. 2013, Honig et al. 2012, Oeseburg et al. 2010, Sanders et al. 2012, Steffens et al. 2012, Weischer et al. 2013a, Wentzensen et al. 2011).

In terms of cancer risks, Wu et al. (Wu et al. 2003) published the first case control study that evaluated the association of LTL with the risk of cancers. In this pioneering study, mean LTL from patients of lung, bladder, head and neck, and renal cancers was significantly shorter than controls, supporting that short LTL was associated with increased cancer risks. In addition, LTL was statistically significantly and inversely associated with baseline and mutagen-induced mutagen sensitivity, indicating short LTL predisposes to cancer through higher genetic instability. Since this paper, numerous epidemiological studies have evaluated the association of LTL with the risk of many different cancers (Hou et al. 2012, Prescott et al. 2012, Wentzensen et al. 2011). The initial hypothesis was that short LTL would predispose to genetic instability, hence increased cancer risks. The majority of retrospective case control studies and a few prospective studies have supported this hypothesis (Hou et al. 2012, Prescott et al. 2012, Wentzensen et al. 2011). Meta-analyses of earlier studies showed significant associations of LTL with overall cancer risk (Ma et al. 2011, Wentzensen et al. 2011). However, with the emergence of very large, prospective studies, the initial striking effect of LTL on cancer risk have attenuated, and it appears that LTL is modestly associated with altered cancer risks in a cancer type-specific manner: in some cancers, short LTL confers increased risks; in some other cancers, LTL was not associated with risk; and surprisingly, long LTL can also confer increased risk in a few cancers (Gu and Wu 2013, Weischer et al. 2013b). Weischer et al. (Weischer et al. 2013b) published the largest prospective cohort study of LTL and cancer risk. The authors measured LTL in 47,102 Danish general population participants from two Denmark cohorts. Participants were followed for up to 20 years for cancer diagnosis and death. During follow-up, there were 3,142 first cancer occurrences, and, among these cancers, there were 1,730 deaths. Multivariate Cox model showed that short LTL was significantly associated with increased risk of cancer death. However, multivariable-adjusted HRs for overall cancer risks were 0.98 (95% CI, 0.88 to 1.08) and 0.95 (95% CI, 0.80 to 1.11) for individuals in the quartile and decile with the shortest vs. the longest telomeres, respectively, suggesting that LTL was not associated with overall cancer risk, a conclusion contrasts with prior meta-analyses (Ma et al. 2011, Wentzensen et al. 2011). However, looking at the odds ratio (OR) of each specific cancer site, it is clear that the association of short LTL with individual cancer is heterogeneous: both short and long LTL can confer increased cancer risks, depending on cancer types and histologies of the same cancer type.

Short LTL and increased cancer risks

In cancers of bladder, esophagus, and oral cavity, both prospective and retrospective studies showed short LTL increases their risks (Bau et al. 2013, Gu et al. 2011, Ma et al. 2011, Wentzensen et al. 2011). For example, in a case control study, we measured LTL in healthy controls, patients with oral premalignant lesions (OPLs) or oral squamous cell carcinoma (OSCC) and found that the age-adjusted relative LTL was the shortest in OSCC (1.64 ± 0.29), intermediate in OPL (1.75 ± 0.43), and longest in controls (1.82 ± 0.36) (P for trend < 0.001). When dichotomized at the median value in controls, adjusting for age, gender, smoking and alcohol drinking status, the ORs for OPL and OSCC risks associated with short LTL was 2.03 (95% CI = 1.29–3.21) and 3.47 (95% CI = 1.84–6.53), respectively, with significant dose-response effects for both associations. Furthermore, among 174 OPL patients, 23 progressed to OSCC during follow-up and the mean LTL was shorter than in progressors than non-progressors (1.66 ± 0.35 vs. 1.77 ± 0.44), although the difference did not reach statistical significance (P = 0.258) likely due to the small number of progressors. In the prospective study of Weischer et al. (Weischer et al. 2013b), the association of LTL with the risk of oral cancer was borderline significant (HR = 1.17, 95% CI = 0.90–1.53). For bladder cancer, in a large case control study of nearly 1,000 pairs of cases and controls, short LTL (dichotomized at the median level in controls) was associated with a 1.42-fold (95% CI = 1.17–1.71) (Gu et al. 2011), consistent with an earlier prospective, nested case control study in which short LTL (comparing lowest quartile with the longest quartile) was associated with a 1.88-fold (95% CI = 1.05–3.36) increased risk of bladder cancer (McGrath et al. 2007).

Long LTL and increased cancer risks

Recently, there have been increasing studies showing that long LTL is associated with increased risks of several cancers. The most consistent result was from melanoma with several independent studies, both retrospective and prospective, reporting that short LTL was protective and long LTL was associated with increased melanoma risk (Anic et al. 2013, Llorca-Cardenosa et al. 2014, Nan et al. 2011, Weischer et al. 2013b). A recent meta-analysis of eight published papers on the association between LTL and the risk of melanoma found a summary relative risk (SRR) of 0.25 (95% CI 0.09–0.67) for cutaneous melanoma when comparing those in the lowest with the highest category of LTL distribution. Furthermore, Iles et al. (Iles et al. 2014) genotyped or imputed 7 SNPs previously associated with LTL in 11,108 melanoma patients and 13,933 controls. A genetic score that predicts long LTL, derived from these seven LTL-associated SNPs, was strongly associated ($P = 8.92 \times 10^{-9}$, two-sided) with increased melanoma risk, providing the first proof that multiple germline genetic determinants of LTL influence cancer risk. The second example of short LTL being protective and long LTL being risk factor is sarcoma. In the cohort study of Weischer et al. (Weischer et al. 2013b), short LTL was strongly associated with a significantly reduced risk of sarcoma (HR = 0.70, 95% CI, 0.55–0.89, P = 0.003), consistent with an earlier retrospective case control study that showed a very strong protective effect of short LTL on soft tissue sarcoma (P < 0.001) (Xie et al. 2013). There were also reports of long LTL with increased risks of other cancer types, such as non-Hodgkin's lymphoma (Lan et al. 2009), B-cell lymphoma (Hosnijeh et al. 2014), endometrial cancer (Sun et al. 2015, Weischer et al. 2013b), prostate cancer (Julin et al. 2015, Weischer et al. 2013b), and lung cancer (Machiela et al. 2015, Sanchez-Espiridion et al. 2014, Seow et al. 2014, Zhang et al. 2015).

Histology-dependent association of LTL with lung cancer risk

Several recent studies have provided strong evidence for a histology-dependent association between LTL and risk of lung cancer: long LTL was associated with an increased risk of lung adenocarcinoma (AC), whereas short LTL may be associated with an increased risk of lung squamous cell carcinoma (SCC). In a large case controls study of 1,385 patients with lung cancer and 1,385 matched controls, Sanchez-Espiridion et al. (Sanchez-Espiridion et al. 2014) measured LTL and performed stratified analyses by histology. They found that AC patients had significantly longer LTL than controls (mean ± SD, 1.23 ± 0.38 vs. 1.14 ± 0.37; P < 0.001), which was consistent regardless of sex, age, and smoking status. Dichotomized at the median LTL in the controls, individuals with long LTLs exhibited significantly increased risk of lung AC (OR = 1.56, 95% CI, 1.23–1.98). In contrast, SCC patients had shorter LTL than the controls (mean ± SD, 1.10 ± 0.44 vs. 1.13 ± 0.33; P = 0.015). Dichotomized at the median LTL in the controls, individuals with long LTLs exhibited a borderline significantly reduced risk of lung SCC (OR, 0.66, 95% CI, 0.42–1.03; P = 0.068). This was the first study to demonstrate an opposite association between LTL and cancer risk between different histologies. Subsequently, Seow et al. (Seow et al. 2014) reported the association between LTL and lung cancer in a pooled analysis from three prospective cohort studies: the Prostate, Lung, Colorectal and Ovarian (PLCO) Cancer Screening Trial, conducted among men and women in the United States, and previously published data from the Alpha-Tocopherol, Beta-Carotene Cancer Prevention (ATBC) trial conducted among male smokers in Finland, and the Shanghai Women's Health Study (SWHS), which is comprised primarily of never-smokers. The pooled population included 847 cases and 847 controls matched by study, age, and sex. They found that longer LTL was associated with increased lung cancer risk [OR (95% CI) by quartile: 1.00; 1.24 (0.90–1.71); 1.27 (0.91–1.78); and 1.86 (1.33–2.62); P trend = 0.000022]. Findings were consistent across the three cohorts. In stratified analysis by histology, the association was only evident in AC, but not in SCC, consistent with the above case control study. More recently, Zhang et al. (Zhang et al. 2015) estimated associations between nine LTL-associated SNPs and risk for five common cancer types (breast, lung, colorectal, ovarian and prostate cancer, including subtypes) using data on 51,725 cases and 62,035 controls by calculating a weighted genetic score representing long LTL. Consistently with the above studies of direct measurement of LTL, they found that long LTL genetic score was significantly associated with increased risk of lung AC (OR = 2.87, 95% CI = 2.20–3.74, P = 6.3×10^{-15}), but not SCC (OR = 1.04, 95% CI = 0.79–1.36, P = 0.79). Likewise, in a similar large scale genetics-based association study consisting of 5,457 never-smoking female Asian lung cancer cases and 4,493 never-smoking female Asian controls, the authors (Machiela et al. 2015) constructed a genetic risk score (GRS) using seven LTL-associated SNPs to predict longer LTL and found the GRS was associated with increased lung cancer risk (OR = 1.51 (95% CI = 1.34–1.69) for upper vs. lower quartile of the weighted GRS, p value = 4.54×10^{-14}), although in this Asian population, there was no difference in GRS effect between AC and SCC subtypes. These genetics-based association studies avoids issues with reverse causality and residual confounding that arise in observational studies of LTL and cancer risks. The evidence is compelling for a significant association between long LTL and increased risk of lung AC, but not in SCC, in Caucasians. In Asians, there may not be a differential effect.

Chromosome-specific LTL and cancer risk

All the above summarized studies reported the associations of mean LTL of 23 pairs of chromosomes (92 telomeres) with cancer risks. However, individual telomere length on each chromosome is highly heterogeneous with some chromosomes having longer and others having shorter telomeres. For example, chromosome 17p has been consistently shown to be one of the shortest telomeres in all cells and all individuals, suggesting telomere shortening in 17p may contribute to its frequent loss in human cancer (Graakjaer et al. 2003, Martens et al. 1998). Hemann et al. (Hemann et al. 2001) revealed that loss of telomere function occurred preferentially on chromosomes with critically short telomeres and suggested that it is not the average but rather the shortest telomeres that constitute telomere dysfunction and limit cellular survival. It is therefore plausible that telomere shortening in each specific chromosome arm confers differential cancer risk. Due to the technical challenge and/or low-throughput of chromosome-specific telomere length assays, there were only a few exploratory studies investigating the associations of specific telomeres with cancer risks. The first study used a PCR-based single telomere length analysis (STELA) assay (Britt-Compton et al. 2006) to examine overall mean LTL and chromosome-specific LTL of 17p, 12q, 2p, and 11q in a case-control study with 94 esophageal cancer cases and 94 matched controls (Xing et al. 2009). Compared with controls, patients had significantly shorter overall LTL (P = 0.004) and chromosome-specific LTL of 17p (P = 0.003) and 12q (P = 0.006) but not of 11q (P = 0.632) and 2p (P = 0.972). Furthermore, the multivariate logistic regression analysis showed that the short overall telomere length and chromosome-specific telomere lengths of 17p and 12q were associated with a dose-dependent increase in EC risk. This study provides the first epidemiological evidence of the differential effect of specific telomere length on cancer risk. In the second study, Zheng et al. (Zheng et al. 2011) examined the associations between individual lengths of 92 telomeres in the human genome and the risk of breast cancer in 204 newly diagnosed breast cancer patients and 236 healthy controls. Chromosome arm-specific telomere lengths were measured by qFISH. This genome-wide screen identified that shorter telomere lengths on chromosomes Xp and 15p were associated with breast cancer risk in pre-menopausal women, with adjusted ORs of 2.5 (95% CI = 1.3–4.8) and 2.6 (95% CI = 1.3–5.0), respectively. The study also revealed that greater length differences between homologous telomeres on chromosomes 9p, 15p and 15q were associated with breast cancer risk in pre-menopausal women, with ORs of 4.6 (95% CI, 2.3–9.2), 3.1 (95% CI, 1.6–6.0) and 2.8 (95% CI, 1.4–5.4), respectively. These studies can provide important biological insights into the role of telomeres in cancer etiology.

Longitudinal changes of LTL and cancer risk

All the above described studies relating LTL to cancer risks have used a single time measurement of LTL without consideration of longitudinal changes of LTL. Recently, Hou et al. (Hou et al. 2015), for the first time, reported a dynamic change of LTL in the context of cancer risks. This study included 792 Normative Aging Study participants with one to four LTL measurements during a follow-up of 12-years. The authors observed that age-related LTL attrition was accelerated among participants who ultimately developed cancer. Strikingly, this trend reversed when age-adjusted LTL was examined relative to time to diagnosis. This divergence of age-adjusted LTL attrition began seven years pre-diagnosis and culminated in significantly longer LTL 3–4 years pre-diagnosis among cancer cases compared to cancer-free participants, resulting in a positive association between longer LTL measured within 4 years of diagnosis and increased risks of prostate cancer and all cancers combined.

This divergence of LTL at some point before cancer diagnosis has important implications in assessing the relationship of LTL with the cancer risks. Current literature in this regard all measured LTL at a single time point. The timing of blood collection was not strictly controlled. This study (Hou et al. 2015) indicates that the differences in sample collection time relative to cancer development and diagnosis will have significant impact on assessing the association of LTL with cancer risks, which at least partially contributes to the inconsistency in literature. Future association studies of single time measurement of LTL should carefully control the timing of blood collection. More importantly, studies with multiple blood collection are warranted to establish the temporal associations between LTL and cancer risks, and to assess the dynamic changes of LTL in relation to cancer development.

Summary and Perspective

The epidemiology, genetics, and cancer predisposing research of LTL have blossomed in the past decade. The association of LTL with cancer risk appear to be cancer type- and histology-specific. The original hypothesis of short LTL predisposes to cancer development has been challenged and more evidence suggest that long LTL can also predispose to cancer development. The most consistent association of long LTL with increased cancer risks occur to melanoma, sarcoma, and lung adenocarcinoma. Short LTL was associated with increased risks of a few cancers, such as bladder, esophageal, and oral cancers. No association was consistently observed for a few cancers, such as colorectal cancer. However, for most cancer types, the literature is highly inconsistent. For example, among dozens of association studies of LTL and breast cancer risk, negative, positive, and null associations have all been reported (Ma et al. 2011, Weischer et al. 2013b, Wentzensen et al. 2011). For pancreatic cancer, a large retrospective case control study with nearly 500 cases and 1000 controls showed shorter LTL being associated with an increased risk for pancreatic cancer, but extremely long telomeres may also be associated with higher risk (Skinner et al. 2012). A subsequent prospective, nested case control study with 193 cases and 660 controls showed longer LTL was significantly associated with increased pancreatic cancer risk (Lynch et al. 2013). In the afore-mentioned largest prospective cohort study (Weischer et al. 2013b), with 124 pancreatic cases among 47,091 participants, the HR was 1.14 (95% CI, 0.93–1.41) for each 1000 base pairs LTL decrease, indicating short LTL was associated with increased risk. These heterogeneous results reflect the challenges of LTL association studies in terms of technical reproducibility, study design, sample size, population heterogeneity, and sample collection time. Large, prospective studies are clearly desired in LTL-cancer association study. Longitudinal studies with repeated measurement of LTL within the same individual will provide dynamic changes of LTL in relation to cancer development. With improved techniques, optimal study design, large sample size, and advanced statistical tools, we are in a much better position to uncover the role of constitutive telomere length in cancer etiology. Full understanding of LTL epidemiology, genetics, and cancer association will facilitate intervention efforts to prevent cancer and improve our lives through modulating telomere dynamics.

Keywords: Leukocyte telomere length, cancer risk, genetic susceptibility, SNP, cohort, case-control study

References

Andrew, T., A. Aviv, M. Falchi, G.L. Surdulescu, J.P. Gardner, X. Lu et al. 2006. Mapping genetic loci that determine leukocyte telomere length in a large sample of unselected female sibling pairs. Am. J. Hum. Genet. 78: 480–486.

Anic, G.M., V.K. Sondak, J.L. Messina, N.A. Fenske, J.S. Zager, B.S. Cherpelis et al. 2013. Telomere length and risk of melanoma, squamous cell carcinoma, and basal cell carcinoma. Cancer Epidemiol. 37: 434–439.

Aubert, G. and P.M. Lansdorp. 2008. Telomeres and aging. Physiol. Rev. 88: 557–579.

Aubert, G., M. Hills and P.M. Lansdorp. 2012. Telomere length measurement-caveats and a critical assessment of the available technologies and tools. Mutat. Res. 730: 59–67.

Aviv, A., A.M. Valdes and T.D. Spector. 2006. Human telomere biology: pitfalls of moving from the laboratory to epidemiology. Int. J. Epidemiol. 35: 1424–1429.

Aviv, A., S.C. Hunt, J. Lin, X. Cao, M. Kimura and E. Blackburn. 2011. Impartial comparative analysis of measurement of leukocyte telomere length/DNA content by Southern blots and qPCR. Nucleic Acids Res. 39: e134.

Bau, D.T., S.M. Lippman, E. Xu, Y. Gong, J.J. Lee, X. Wu et al. 2013. Short telomere lengths in peripheral blood leukocytes are associated with an increased risk of oral premalignant lesion and oral squamous cell carcinoma. Cancer. 119: 4277–4283.

Blackburn, E.H. 2005. Telomeres and telomerase: their mechanisms of action and the effects of altering their functions. FEBS Lett. 579: 859–862.

Blackburn, E.H., C.W. Greider and J.W. Szostak. 2006. Telomeres and telomerase: the path from maize, Tetrahymena and yeast to human cancer and aging. Nat. Med. 12: 1133–1138.

Bojesen, S.E., K.A. Pooley, S.E. Johnatty, J. Beesley, K. Michailidou, J.P. Tyrer et al. 2013. Multiple independent variants at the TERT locus are associated with telomere length and risks of breast and ovarian cancer. Nat. Genet. 45: 371–384, 384e371–372.

Britt-Compton, B., J. Rowson, M. Locke, I. Mackenzie, D. Kipling and D.M. Baird. 2006. Structural stability and chromosome-specific telomere length is governed by cis-acting determinants in humans. Human Molecular Genetics. 15: 725–733.

Broer, L., V. Codd, D.R. Nyholt, J. Deelen, M. Mangino, G. Willemsen et al. 2013. Meta-analysis of telomere length in 19 713 subjects reveals high heritability, stronger maternal inheritance and a paternal age effect. European Journal of Human Genetics: EJHG.

Brouilette, S.W., J.S. Moore, A.D. McMahon, J.R. Thompson, I. Ford, J. Shepherd et al. 2007. Telomere length, risk of coronary heart disease, and statin treatment in the West of Scotland primary prevention study: a nested case-control study. Lancet. 369: 107–114.

Bruderlein, S., K. Muller, J. Melzner, J. Hogel, P. Wiegand and P. Moller. 2008. Different rates of telomere attrition in peripheral lymphocytes in a pair of dizygotic twins with hematopoietic chimerism. Aging Cell. 7: 663–666.

Cawthon, R.M. 2002. Telomere measurement by quantitative PCR. Nucleic Acids Res. 30: e47.

Codd, V., M. Mangino, P. van der Harst, P.S. Braund, M. Kaiser, A.J. Beveridge et al. 2010. Common variants near TERC are associated with mean telomere length. Nat. Genet. 42: 197–199.

Codd, V., C.P. Nelson, E. Albrecht, M. Mangino, J. Deelen, J.L. Buxton et al. 2013. Identification of seven loci affecting mean telomere length and their association with disease. Nature Genetics. 45: 422–427, 427e421–422.

Du, J., X. Zhu, C. Xie, N. Dai, Y. Gu, M. Zhu et al. 2015. Telomere length, genetic variants and gastric cancer risk in a Chinese population. Carcinogenesis. 36: 963–970.

Fyhrquist, F., O. Saijonmaa and T. Strandberg. 2013. The roles of senescence and telomere shortening in cardiovascular disease. Nat. Rev. Cardiol. 10: 274–283.

Graakjaer, J., C. Bischoff, L. Korsholm, S. Holstebroe, W. Vach, V.A. Bohr et al. 2003. The pattern of chromosome-specific variations in telomere length in humans is determined by inherited, telomere-near factors and is maintained throughout life. Mech. Ageing Dev. 124: 629–640.

Graakjaer, J., H. Der-Sarkissian, A. Schmitz, J. Bayer, G. Thomas, S. Kolvraa et al. 2006a. Allele-specific relative telomere lengths are inherited. Hum. Genet. 119: 344–350.

Graakjaer, J., J.A. Londono-Vallejo, K. Christensen and S. Kolvraa. 2006b. The pattern of chromosome-specific variations in telomere length in humans shows signs of heritability and is maintained through life. Ann. N Y Acad. Sci. 1067: 311–316.

Gu, J., M. Chen, S. Shete, C.I. Amos, A. Kamat, Y. Ye et al. 2011. A genome-wide association study identifies a locus on chromosome 14q21 as a predictor of leukocyte telomere length and as a marker of susceptibility for bladder cancer. Cancer Prev. Res. (Phila). 4: 514–521.

Gu, J. and X. Wu. 2013. Re: short telomere length, cancer survival, and cancer risk in 47 102 individuals. J. Natl. Cancer Inst. 105: 1157.

Hemann, M.T., M.A. Strong, L.Y. Hao and C.W. Greider. 2001. The shortest telomere, not average telomere length, is critical for cell viability and chromosome stability. Cell. 107: 67–77.

Honig, L.S., M.S. Kang, N. Schupf, J.H. Lee and R. Mayeux. 2012. Association of shorter leukocyte telomere repeat length with dementia and mortality. Arch. Neurol. 69: 1332–1339.

Hosnijeh, F.S., G. Matullo, A. Russo, S. Guarrera, F. Modica, A. Nieters et al. 2014. Prediagnostic telomere length and risk of B-cell lymphoma-Results from the EPIC cohort study. International Journal of Cancer. Journal International du Cancer. 135: 2910–2917.

Hou, L., X. Zhang, A.J. Gawron and J. Liu. 2012. Surrogate tissue telomere length and cancer risk: shorter or longer? Cancer Letters. 319: 130–135.

Hou, L., B.J. Joycea, T. Gaoa, L. Liu, Y. Zheng, F.J. Penedo et al. 2015. Blood Telomere Length Attrition and Cancer Development in the Normative Aging Study Cohort. EBioMedicine.

Iles, M.M., D.T. Bishop, J.C. Taylor, N.K. Hayward, M. Brossard, A.E. Cust et al. 2014. The effect on melanoma risk of genes previously associated with telomere length. J. Natl. Cancer Inst. 106.

Jeanclos, E., N.J. Schork, K.O. Kyvik, M. Kimura, J.H. Skurnick and A. Aviv. 2000. Telomere length inversely correlates with pulse pressure and is highly familial. Hypertension. 36: 195–200.

Julin, B., I. Shui, C.M. Heaphy, C.E. Joshu, A.K. Meeker, E. Giovannucci et al. 2015. Circulating leukocyte telomere length and risk of overall and aggressive prostate cancer. British Journal of Cancer. 112: 769–776.

Kim, N.W., M.A. Piatyszek, K.R. Prowse, C.B. Harley, M.D. West, P.L. Ho et al. 1994. Specific association of human telomerase activity with immortal cells and cancer. Science. 266: 2011–2015.

Kimura, M., R.C. Stone, S.C. Hunt, J. Skurnick, X. Lu, X. Cao et al. 2010. Measurement of telomere length by the Southern blot analysis of terminal restriction fragment lengths. Nat. Protoc. 5: 1596–1607.

Lan, Q., R. Cawthon, M. Shen, S.J. Weinstein, J. Virtamo, U. Lim et al. 2009. A prospective study of telomere length measured by monochrome multiplex quantitative PCR and risk of non-Hodgkin lymphoma. Clinical Cancer Research: An Official Journal of the American Association for Cancer Research. 15: 7429–7433.

Lansdorp, P.M., N.P. Verwoerd, F.M. van de Rijke, V. Dragowska, M.T. Little, R.W. Dirks et al. 1996. Heterogeneity in telomere length of human chromosomes. Hum. Mol. Genet. 5: 685–691.

Levy, D., S.L. Neuhausen, S.C. Hunt, M. Kimura, S.J. Hwang, W. Chen et al. 2010. Genome-wide association identifies OBFC1 as a locus involved in human leukocyte telomere biology. Proc. Natl. Acad. Sci. USA. 107: 9293–9298.

Lin, J., E. Epel and E. Blackburn. 2012. Telomeres and lifestyle factors: roles in cellular aging. Mutat. Res. 730: 85–89.

Llorca-Cardenosa, M.J., M. Pena-Chilet, M. Mayor, C. Gomez-Fernandez, B. Casado, M. Martin-Gonzalez et al. 2014. Long telomere length and a TERT-CLPTM1 locus polymorphism association with melanoma risk. European Journal of Cancer. 50: 3168–3177.

Ludlow, A.T. and S.M. Roth. 2011. Physical activity and telomere biology: exploring the link with aging-related disease prevention. J. Aging Res. 2011: 790378.

Lynch, S.M., J.M. Major, R. Cawthon, S.J. Weinstein, J. Virtamo, Q. Lan et al. 2013. A prospective analysis of telomere length and pancreatic cancer in the alpha-tocopherol beta-carotene cancer (ATBC) prevention study. International Journal of Cancer. Journal International du Cancer.

Ma, H., Z. Zhou, S. Wei, Z. Liu, K.A. Pooley, A.M. Dunning et al. 2011. Shortened telomere length is associated with increased risk of cancer: a meta-analysis. PloS One. 6: e20466.

Machiela, M.J., C.A. Hsiung, X.O. Shu, W.J. Seow, Z. Wang, K. Matsuo et al. 2015. Genetic variants associated with longer telomere length are associated with increased lung cancer risk among never-smoking women in Asia: a report from the female lung cancer consortium in Asia. International Journal of Cancer. Journal International du Cancer. 137: 311–319.

Mangino, M., S. Brouilette, P. Braund, N. Tirmizi, M. Vasa-Nicotera, J.R. Thompson et al. 2008. A regulatory SNP of the BICD1 gene contributes to telomere length variation in humans. Hum. Mol. Genet. 17: 2518–2523.

Mangino, M., S.J. Hwang, T.D. Spector, S.C. Hunt, M. Kimura, A.L. Fitzpatrick et al. 2012. Genome-wide meta-analysis points to CTC1 and ZNF676 as genes regulating telomere homeostasis in humans. Hum. Mol. Genet. 21: 5385–5394.

Mangino, M., L. Christiansen, R. Stone, S.C. Hunt, K. Horvath, D.T. Eisenberg et al. 2015. DCAF4, a novel gene associated with leucocyte telomere length. J. Med. Genet. 52: 157–162.

Martens, U.M., J.M. Zijlmans, S.S. Poon, W. Dragowska, J. Yui, E.A. Chavez et al. 1998. Short telomeres on human chromosome 17p. Nat. Genet. 18: 76–80.

McGrath, M., J.Y. Wong, D. Michaud, D.J. Hunter and I. De Vivo. 2007. Telomere length, cigarette smoking, and bladder cancer risk in men and women. Cancer Epidemiol. Biomarkers Prev. 16: 815–819.

Nan, H., M. Du, I. De Vivo, J.E. Manson, S. Liu, A. McTiernan et al. 2011. Shorter telomeres associate with a reduced risk of melanoma development. Cancer Res. 71: 6758–6763.

Needham, B.L., N. Adler, S. Gregorich, D. Rehkopf, J. Lin, E.H. Blackburn et al. 2013. Socioeconomic status, health behavior, and leukocyte telomere length in the National Health and Nutrition Examination Survey, 1999–2002. Soc. Sci. Med. 85: 1–8.

Nettleton, J.A., A. Diez-Roux, N.S. Jenny, A.L. Fitzpatrick and D.R. Jr. Jacobs. 2008. Dietary patterns, food groups, and telomere length in the Multi-Ethnic Study of Atherosclerosis (MESA). Am. J. Clin. Nutr. 88: 1405–1412.

O'Sullivan, J., R.A. Risques, M.T. Mandelson, L. Chen, T.A. Brentnall, M.P. Bronner et al. 2006. Telomere length in the colon declines with age: a relation to colorectal cancer? Cancer Epidemiol. Biomarkers Prev. 15: 573–577.

Oeseburg, H., R.A. de Boer, W.H. van Gilst and P. van der Harst. 2010. Telomere biology in healthy aging and disease. Pflugers Arch. 459: 259–268.

Pooley, K.A., M.S. Sandhu, J. Tyrer, M. Shah, K.E. Driver, R.N. Luben et al. 2010. Telomere length in prospective and retrospective cancer case-control studies. Cancer Res. 70: 3170–3176.

Pooley, K.A., S.E. Bojesen, M. Weischer, S.F. Nielsen, D. Thompson, A. Amin Al Olama et al. 2013. A genome-wide association scan (GWAS) for mean telomere length within the COGS project: identified loci show little association with hormone-related cancer risk. Hum. Mol. Genet.

Prescott, J., I.M. Wentzensen, S.A. Savage and I. De Vivo. 2012. Epidemiologic evidence for a role of telomere dysfunction in cancer etiology. Mutat. Res. 730: 75–84.

Price, L.H., H.T. Kao, D.E. Burgers, L.L. Carpenter and A.R. Tyrka. 2013. Telomeres and early-life stress: an overview. Biol. Psychiatry. 73: 15–23.

Sanchez-Espiridion, B., M. Chen, J.Y. Chang, C. Lu, D.W. Chang, J.A. Roth et al. 2014. Telomere length in peripheral blood leukocytes and lung cancer risk: a large case-control study in Caucasians. Cancer Res. 74: 2476–2486.

Sanders, J.L., A.L. Fitzpatrick, R.M. Boudreau, A.M. Arnold, A. Aviv, M. Kimura et al. 2012. Leukocyte telomere length is associated with noninvasively measured age-related disease: The Cardiovascular Health Study. The Journals of Gerontology. Series A, Biological Sciences and Medical Sciences. 67: 409–416.

Sanders, J.L. and A.B. Newman. 2013. Telomere Length in Epidemiology: A Biomarker of Aging, Age-Related Disease, Both, or Neither? Epidemiol Rev.

Seow, W.J., R.M. Cawthon, M.P. Purdue, W. Hu, Y.T. Gao, W.Y. Huang et al. 2014. Telomere length in white blood cell DNA and lung cancer: a pooled analysis of three prospective cohorts. Cancer Res. 74: 4090–4098.

Shi, J., F. Sun, L. Peng, B. Li, L. Liu, C. Zhou et al. 2013. Leukocyte telomere length-related genetic variants in 1p34.2 and 14q21 loci contribute to the risk of esophageal squamous cell carcinoma. International Journal of Cancer. Journal International du Cancer. 132: 2799–2807.

Skinner, H.G., R.E. Gangnon, K. Litzelman, R.A. Johnson, S.T. Chari, G.M. Petersen et al. 2012. Telomere length and pancreatic cancer: a case-control study. Cancer Epidemiol. Biomarkers Prev. 21: 2095–2100.

Slagboom, P.E., S. Droog and D.I. Boomsma. 1994. Genetic determination of telomere size in humans: a twin study of three age groups. Am. J. Hum. Genet. 55: 876–882.

Steffens, J.P., S. Masi, F. D'Aiuto and L.C. Spolidorio. 2012. Telomere length and its relationship with chronic diseases—New perspectives for periodontal research. Arch. Oral. Biol.

Sun, Y., L. Zhang, L. Zhao, X. Wu and J. Gu. 2015. Association of leukocyte telomere length in peripheral blood leukocytes with endometrial cancer risk in Caucasian Americans. Carcinogenesis. 36: 1327–1332.

Valdes, A.M., T. Andrew, J.P. Gardner, M. Kimura, E. Oelsner, L.F. Cherkas et al. 2005. Obesity, cigarette smoking, and telomere length in women. Lancet. 366: 662–664.

Vasa-Nicotera, M., S. Brouilette, M. Mangino, J.R. Thompson, P. Braund, J.R. Clemitson et al. 2005. Mapping of a major locus that determines telomere length in humans. Am. J. Hum. Genet. 76: 147–151.

Walsh, K.M., V. Codd, I.V. Smirnov, T. Rice, P.A. Decker, H.M. Hansen et al. 2014. Variants near TERT and TERC influencing telomere length are associated with high-grade glioma risk. Nat. Genet. 46: 731–735.

Weischer, M., B.G. Nordestgaard, R.M. Cawthon, J.J. Freiberg, A. Tybjaerg-Hansen and S.E. Bojesen. 2013a. Short telomere length, cancer survival, and cancer risk in 47102 individuals. J. Natl. Cancer Inst. 105: 459–468.

Weischer, M., B.G. Nordestgaard, R.M. Cawthon, J.J. Freiberg, A. Tybjaerg-Hansen and S.E. Bojesen. 2013b. Short telomere length, cancer survival, and cancer risk in 47102 individuals. J. Natl. Cancer Inst. 105: 459–468.

Wentzensen, I.M., L. Mirabello, R.M. Pfeiffer and S.A. Savage. 2011. The association of telomere length and cancer: a meta-analysis. Cancer Epidemiol. Biomarkers Prev. 20: 1238–1250.

Wu, X., C.I. Amos, Y. Zhu, H. Zhao, B.H. Grossman, J.W. Shay et al. 2003. Telomere dysfunction: a potential cancer predisposition factor. J. Natl. Cancer Inst. 95: 1211–1218.

Xie, H., X. Wu, S. Wang, D. Chang, R.E. Pollock, D. Lev et al. 2013. Long telomeres in peripheral blood leukocytes are associated with an increased risk of soft tissue sarcoma. Cancer. 119: 1885–1891.

Xing, J., J.A. Ajani, M. Chen, J. Izzo, J. Lin, Z. Chen et al. 2009. Constitutive short telomere length of chromosome 17p and 12q but not 11q and 2p is associated with an increased risk for esophageal cancer. Cancer Prev. Res. (Phila). 2: 459–465.

Zhang, C., J.A. Doherty, S. Burgess, R.J. Hung, S. Lindstrom, P. Kraft et al. 2015. Genetic determinants of telomere length and risk of common cancers: a Mendelian randomization study. Hum. Mol. Genet. 24: 5356–5366.

Zheng, Y.L., X. Zhou, C.A. Loffredo, P.G. Shields and B. Sun. 2011. Telomere deficiencies on chromosomes 9p, 15p, 15q and Xp: potential biomarkers for breast cancer risk. Hum. Mol. Genet. 20: 378–386.

Chapter 5

Telomere Length in Cardiovascular Disease

Marilena Giannoudi[1] and *Ioakim Spyridopoulos*[2,*]

INTRODUCTION

This chapter examines the role of leukocyte telomere length (LTL) as a marker of cardiovascular ageing. Different techniques to quantify telomere length will be discussed. We will explore a meta-analysis of all relevant studies and discuss the potential link between LTL and coronary artery disease (CAD). Despite the well documented fact that CAD is associated with short LTL, there has been no proof of causality until now. Short telomere length in peripheral blood leukocytes itself can have an inherited component without the co-existence of DNA damage, indicate increased cell turnover, e.g., under systemic low grade inflammation, or be a surrogate for accelerated DNA damage and cell senescence. Future Mendelian randomization studies could provide a valuable tool to look into a mechanistic link between LTL and CAD.

Telomeres are located at the end of each chromosome, protecting the integrity of the mammalian DNA. In non-germline and non-cancerous cells, they shorten with each cell division (Fig. 1). Cell proliferation occurs at every stage of our lives, whether in the bone marrow, the immune system, in skin cells, arterial wall, or, at a slower rate, in the heart. Together with the assumption that certain diseases can accelerate this process, scientists have proposed that telomeres can be viewed as the 'clock of life' (Vega et al. 2003, Blasco 2005). By measuring their length in specific cell populations, one could determine the biological rather than chronological age of the organism, or of specific organ systems such as the cardiovascular system. Several studies have also demonstrated the reverse effect, in that short telomeres

[1] Medical School, Newcastle University, Newcastle upon Tyne, UK.
[2] Institute of Genetic Medicine, Newcastle University, Central Parkway, Newcastle Upon Tyne, NE1 3BZ, UK.
* Corresponding author: ioakim.spyridopoulos@newcastle.ac.uk

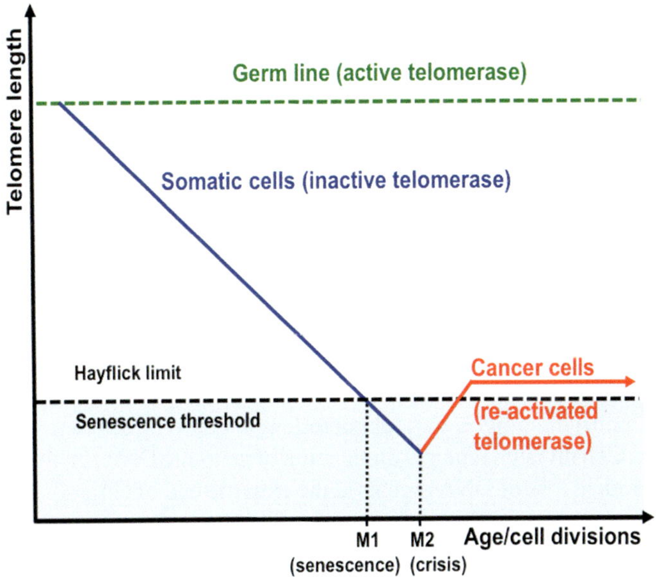

Fig. 1. Model for telomere shortening and senescence. Replicative senescence occurs as a consequence of probably over hundred cell divisions, leading to short telomeres over the lifespan of the cell. The so-called Hayflick limit refers to the observation that cells in culture stop proliferating at some point, co-inciding with short telomeres at his point. This phenomenon applies generally to non-germ line cells as well as non-cancer cells. The latter have the ability to re-activate telomerase, elongate telomeres and thus escape senescence. In contrast to embryonic stem cells, adult stem cells do not have the ability to overcome telomere shortening, reflected by TL in bone marrow cells (Harley et al. 1990).

can themselves cause disease. These propositions are complicated by the fact that telomere length varies dramatically at birth. Here we will explain the underlying biology of telomeres and their molecular regulation. We will subsequently summarize all relevant clinical studies examining a connection between cardiovascular disease and telomere length. Finally, we will attempt to unravel whether there are potential molecular links between TL shortening and cardiovascular disease.

What are Telomeres ?

Telomeres are end-chromosomal caps composed primarily of double-stranded DNA along with some single-stranded DNA. Their role is to protect each chromosome from DNA damage and to maintain the chromatin conformation. The ends of linear chromosomes need to be distinguished from broken DNA ends requiring repair. In the absence of such a distinction, linear chromosomes are prone to illicit DNA end-joining and recombinational events resulting in deleterious consequences such as end-to-end chromosomal fusions. The double-stranded part is composed of multiple repeats of TTAGGG nucleotides on the leading strand (extending from the 5' to the 3' direction towards the end of the chromosome), and their complementary CCCTAA repeats on the other strand (from 3' to 5' direction). In humans, these repeats are

2–15 kB long. At the very end of the telomere, DNA becomes single-stranded and is termed the overhang or tail. This part is approximately 100–300 bp long (Wright et al. 1997, McElligott and Wellinger 1997). It is capped by a protein complex known as shelterin, which is composed of six proteins: TRF1, TRF2, RAP1, TIN2, POT1 and TPP1. The single-stranded overhang serves either as a template for the DNA polymerase telomerase or forms t-loops and thus protects the chromosome end. Removal of any of the 6 shelterin proteins leads to activation of DNA-damage response pathways. Telomeric DNA repeats can also be found distant from the end of chromosomes, where they interact with TRF2 to form interstitial DNA loops, supporting the tertiary structure of the chromosomes.

Why are Telomeres Getting Shorter?

The general understanding is that telomere length can only shorten when cells are proliferating. Cell division requires duplication of genomic DNA, including telomeric DNA. The gradual loss of DNA repeats at the extreme end of chromosomes owing to incomplete replication by DNA polymerases (which only synthesize DNA in the 5' to 3' direction and are unable to fill in the gap left behind by the 5' most RNA primer) defines the chromosome end-replication problem. Oxidative stress has evolved as an important contributory factor in telomere attrition in fibroblasts and endothelial cells (von Zglinicki 2002). Telomeres acquire oxidative single-strand DNA damage faster than the bulk of the genome for three reasons: Firstly, DNA sequences which contain guanine triplets (as do the telomere repeat sequences) are highly sensitive to oxidative modification (von Zglinicki et al. 2000, Saretzki et al. 2003, Oikawa et al. 2001). Secondly, repair of single strand DNA breaks is significantly less efficient in telomeres than in the bulk of the genome (Petersen et al. 1998). Finally, oxidative stress also interferes with telomere maintenance via its suppressive effect on telomerase (see below) (Jakob et al. 2008). Correlative evidence from human population studies collectively suggests that cardiovascular risk factors such as smoking, hypertension and obesity, as well as clinically overt coronary artery disease (CAD) and heart failure, are associated with short telomeres (Epel et al. 2004, Valdes et al. 2005, Samani et al. 2001, Farzaneh-Far et al. 2008, Brouilette et al. 2003, van der Harst et al. 2007). In summary, telomere length in a specific tissue or cell type will depend on 4 factors; the starting telomere length, the rate of proliferation of the cell type, the amount of DNA base pairs lost per cell division due to the end-replication problem, and finally telomerase activity.

Telomerase

Telomere shortening can be compensated for or decelerated by concomitant activation of telomerase, a DNA polymerase and specialized ribonucleoprotein, which is composed of an RNA subunit (TERC) serving as a template for sequence addition, and a reverse transcriptase (TERT) subunit facilitating the replication of telomeres. Most human somatic cells do not express telomerase at detectable levels. In the peripheral blood, B- and T-lymphocytes are the only cells to express significant telomerase activity. Overexpression of the hTERT subunit in fibroblasts and other cell types leads to immortalization, also overcoming shortening of telomeres, and in some cases leading to telomere elongation. In contrast, human hematopoietic cells can transiently up-regulate telomerase activity during proliferation after stimulation with cytokines (Chiu et al. 1996, Wang et al. 2005). However, telomerase activity in hematopoietic stem cells (HSC) is unable to prevent age- and proliferation-associated loss of telomeric DNA (Baerlocher et al. 2003). Overexpression of telomerase in human hematopoietic cells does not prevent proliferation-associated telomere shortening, pointing to the existence of cell type-specific differences in telomere dynamics (Wang et al. 2005).

Telomere Length in Cardiovascular Disease

Numerous studies, conducted in a range of patient populations, link short telomere length (TL) to a range of diseases, including cardiovascular disease and many of its attendant risk factors, such as smoking, diabetes, hypertension, dyslipidemia, psychological stress and high body mass index (Brouilette et al. 2003, Samani et al. 2001, Spyridopoulos et al. 2009, van der Harst et al. 2007, Valdes et al. 2005). Although prospective studies identify short leukocyte telomere length (LTL) as a risk factor in the development of coronary heart disease (Brouilette et al. 2007), proof for a causative link between LTL and CAD in humans has not currently been established, and results from mouse models do not uphold such a link (Poch et al. 2004). Observationally, there is a correlation between shorter telomeres and increased risk of coronary heart disease, independent of the conventional vascular risk factors.

In a substudy of the MERIT-HF study of over 800 patients with **chronic heart failure**, LTL was found to be significantly shorter than in age-matched subjects from the healthy control group (van der Harst et al. 2007). Furthermore, the presence of CAD resulted in further shortening of LTL. Accordingly, when using the number of atherosclerotic manifestations as a score, LTL was found to be shorter for each increase in manifestations. Since chronic heart failure (CHF) is associated with increased systemic oxidative stress and inflammation (Landmesser et al. 2002, Anand et al. 2005), these findings uphold the proposition that heightened inflammation, irrespective of its origin, gives rise to accelerated telomere shortening.

A recent **meta-analysis** of 24 different trials, compiling data from over 40,000 participants and 8,000 patients with cardiovascular disease (Haycock et al. 2014), found the association to be broadly consistent across all the study level characteristics tested. The association was retained across clinically relevant subgroupings, such as variation in mean age and sex distribution. This association remained consistent across both prospective and retrospective studies. The pooled relative risk for coronary artery disease was 1.54 (95% confidence interval 1.30 to 1.83) in all studies and 1.40 (1.15 to 1.70) in prospective studies alone. The association between shorter telomeres and cerebrovascular disease was comparable to the association with coronary heart disease, but was not significant when restricting data solely to prospective studies.

Methods to Measure Telomere Length in Population-based Studies

In this chapter we will briefly explain the main methods currently used in population studies to quantify LTL. When interpreting the results of the numerous clinical studies on telomere length, it is important to comprehend the specific limitations of each method. It is also necessary to note that in most studies, the main source for quantification of TL is DNA extracted from peripheral blood leukocytes, which is by definition a heterogenous source. Some longitudinal studies claim to find telomere length elongation over time (Farzaneh-Far et al. 2010), and this can probably be attributed either to wide variability in TL measurement, or to changes in leukocyte composition at different times in life. From a cell biology point of view, only the overexpression of telomerase, for example in cancer cells, can lead to telomere elongation. Cell division almost always results in telomere shortening.

So far, three main methods have been employed in the quantification of telomere length in cardiovascular studies: Southern blotting (which is regarded as the gold standard), quantitative PCR-based technique, and Flow-FISH. Until now absolute quantification of telomere length has not been possible by any method. A recent international collaborative study found significant variation in LTL based on the same physical DNA samples, which probably stems from the diversity of individual laboratory set-ups (Martin-Ruiz et al. 2014). Furthermore, there is no synthetic standard for calibration of absolute lengths across laboratories. However, the authors found a strong correlation (0.63–0.99) for ranking TL from shortest to longest across 180 samples.

Terminal restriction fragment (TRF) analysis uses digested genomic DNA which is subjected to Southern hybridization with a labeled $(C_3TA_2)_3$ telomeric probe followed by densitometric analysis (Kimura et al. 2010). Despite being relatively time consuming and requiring larger amounts of DNA than the PCR method, this remains the gold standard for the measurement of telomere length (Southern 1975). Since the distance between terminal restriction sites and the start of telomere sequences varies between chromosomes, there is potential for overestimation of true telomere size (Rufer et al. 1998). The coefficient variation (CV) of this method can be as low as 2.5%.

Because of the formation of primer dimers due the TTAGGG nucleotide repeats in telomeres, the **qPCR** method with modified telomere-specific primers is probably the most sensitive method available for quantification of telomere length (Cawthon 2002, Cawthon 2009). It is based on the relative ratio of the telomere-repeat copy number (T) to a single gene copy number (S); the so-called T/S ratio. When T/S ratios are compared in the same experimental setup, the inclusion of a reference DNA sample of known length allows transformation of the ratio into an actual nucleotide length. The CV of this method can be reduced below 3% when using pipetting robots as well as different DNA standard dilutions on each plate. Interestingly, the choice of single-copy gene does not have an impact on CV (Martin-Ruiz et al. 2014). However, in our experience this CV is only low when using the same physical DNA for inter-assay variability. Separate DNA extraction could result in further variability due to residual salt particles impacting the PCR reaction itself. Nevertheless, this has become the method of choice for population-based studies due to its dynamic range, speed and potential for high throughput.

The **Flow-FISH** technique combines the linear hybridization of telomeres with a fluorochorome-marked $(C_3TA_2)_3$-PNA probe and quantification by flow cytometry (Baerlocher et al. 2006). By adding heat-resistant fluorochromes that are conjugated to antibodies, simultaneous immunophenotyping is possible (Multicolor flow-FISH). This produces a rapid and very sensitive method of measuring mean telomere length in different subpopulations of nucleated blood cells. It also provides a huge advantage over all other methods when cell types cannot be separately analysed unless sorted prior to DNA extraction. Reproducibility of experiments using the Flow-FISH protocol is reliant upon four factors; accurate measurement of comparatively weak flourescence signals, precise sample-handling at each protocol step, correct instrument calibration, and inclusion of internal controls (bovine thymocytes). In our hands, this method has a CV of 5%, which can be reduced further by implementing machine-automated washing steps (Spyridopoulos et al. 2009).

Telomere Length Shortening from Birth to Ageing

Most population-based studies on telomere length use peripheral blood leukocytes as their source for quantification. The largest fraction of leukocytes are neutrophil granulocytes, which reflect TL of myeloid-derived stem- and progenitor cells (Fig. 2) (Spyridopoulos et al. 2009). When mononuclear cells are isolated from the peripheral blood (PBMCs), the majority of cells would be lymphocytes, primarily T lymphocytes. Since only the Flow-FISH method or prior cell sorting can determine TL separately for these populations, it is necessary to assume that the majority of studies reflect myeloid TL. However, because lymphocytes shorten more rapidly during life, a large amount of short-telomeric T-cells can affect leukocyte TL (Figs. 3 and 4). While billions of new blood cells are produced daily from hematopoetic stem cells (HSCs), there is evidence that telomerase activity in HCSs cannot prevent telomere loss associated with age and proliferation in this cell compartment (Harley et al. 1990, Blackburn 1991, Wright et al. 1996). Repopulation of the hematopoietic system in the course of bone marrow transplantation has been used as a model to study telomere dynamics of HSCs *in vivo*. Telomere length in peripheral leukocytes of recipients following allogenic bone marrow transplantation is shorter compared to their respective donors, implying that there is a greater proliferative demand placed on stem cells in the course of hematopoietic reconstitution than during normal hematopoiesis (Notaro et al. 1997). Granulocytes are characterized by a lack of proliferation and a relatively short life span (hours to days). We have shown that granulocytes from both young and old donors display similar rates of telomere length loss when compared to telomere length of myeloid bone marrow cells (Fig. 2) (Spyridopoulos et al. 2009). In their landmark study of 301 individuals, spanning all

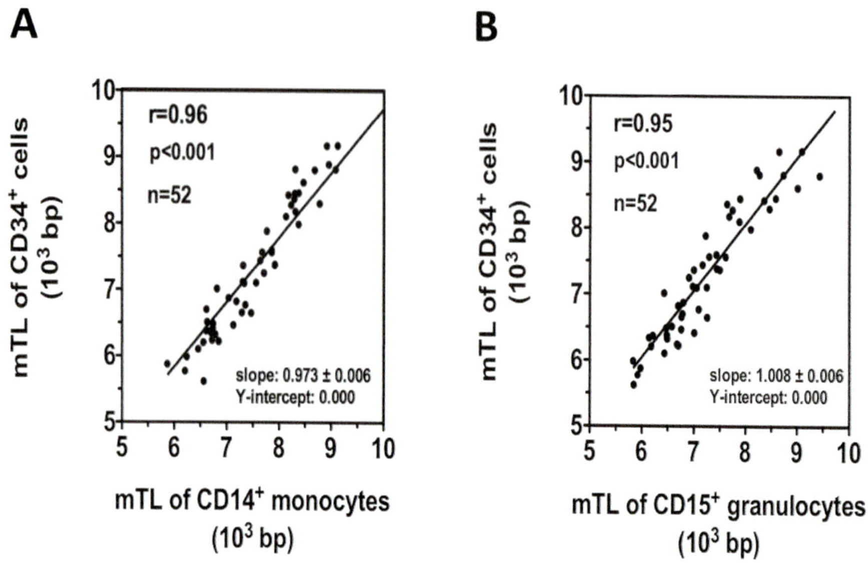

Fig. 2. Telomere length in peripheral blood myeloid cells (granulocytes and monocytes) correlates very well with TL in CD34+ peripheral blood progenitor cells. This justifies using either cell type to predict TL of bone marrow resident progenitor cells (Spyridopoulos et al. 2009).

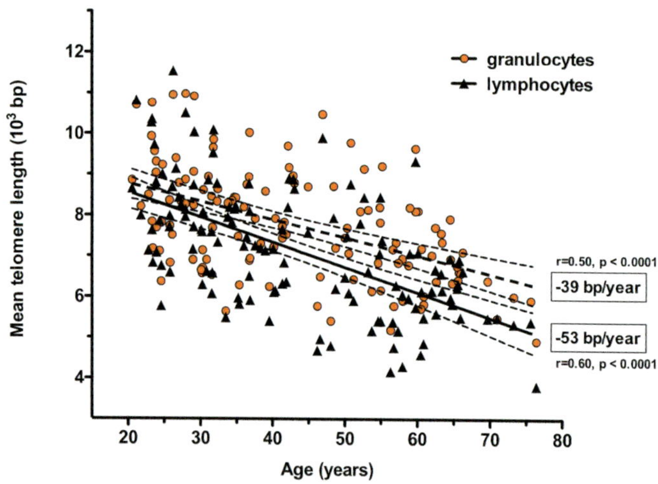

Fig. 3. Age-related telomere shortening in granulocytes and lymphocytes. Mean telomere length (TL) for granulocytes (●) and lymphocytes (▲) is plotted against age. Linear regression curves for the granulocyte and lymphocyte population are shown. Slopes for the age-dependent decline in telomere length were 39 bp/year for the granulocyte population compared to 53 bp/year for the lymphocyte (Spyridopoulos et al. 2008).

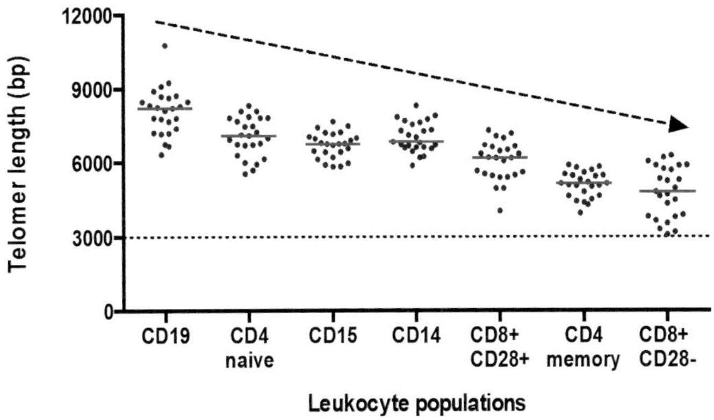

Fig. 4. Telomere length distribution of leukocyte subpopulations. Telomere length (mean ± SEM) was measured in 7 different leukocyte subpopulations from peripheral blood of n = 25 patients with CAD and previous myocardial infarction (mean age 65 years). The data show that in this relatively confined age-range of male patients telomere length did not exhibit a very high standard error in each cell population, despite inherited telomere length differences. B-lymphocytes ($CD19^+$ cells) contain the longest telomeres, followed by naïve CD4 lymphocytes ($CD4^+CD45^{RA+}$), monocytes ($CD14^+$ cells), and granulocytes ($CD15^+$ cells). Cytotoxic ($CD8^+$) T-cells display shorter telomeres, especially memory T-cell subpopulations which lack the CD28 coreceptor ($CD28^-$) (Spyridopoulos et al. 2009).

possible ages from newborns to nonagenarians, Rufer and coworkers demonstrated that telomere shortening featured by granulocytes and lymphocytes *in vivo* follows a cubic function during ageing, which is marked by a significant drop during the first year of life followed by a slower, steady decline thereafter (Rufer et al. 1999). Using the Flow-FISH method, telomere length in granulocytes has been shown to decrease with a rate of 39 bp/year (Rufer et al. 1999). We and others have shown that shortening of telomeres in total lymphocytes occurs significantly faster than it does for telomeres in granulocytes (53 bp/year vs. 39 bp/year respectively, Fig. 3) (Rufer et al. 1999, Spyridopoulos et al. 2008, Spyridopoulos et al. 2009). This difference may be related to the compartment of memory lymphocytes. Since both CD4+ and CD8+ memory T cells contain much shorter telomeres than their naive counterparts (Rufer et al. 1999), the increasing number of effector and memory T-cell subsets with age contributes more swiftly to the decrease in lymphocyte TL (Koch et al. 2008). Granulocytes on the other hand, have been shown to shorten with age at the same rate as naive T cells (Rufer et al. 1999).

Heritability of Telomere length

A recent meta-analysis of almost 20,000 subjects has examined telomere length in relation to heritability, mode of inheritance and parental age at birth (Broer et al. 2013). Three cohorts of twin data were included in the study, with the correlation for monozygotic twins found to be strongest ($r = 0.72$), while dizygotic twins ($r = 0.50$) and siblings ($r = 0.49$) had a much lower but similar correlation. Spouses over the age of 55 also showed a significant correlation ($r = 0.31$), suggesting environmental factors. Given the high heritability, the authors also discovered a stronger maternal than paternal inheritance, as well as a positive association with paternal age.

Predictors of Telomere Length in Genome-wide Association Studies

Genome-wide association studies (GWAS) of 40,000 individuals with replication of selected variants in a further 11,000 individuals identified seven loci, including five novel loci, associated with mean LTL ($P < 5 \times 10^{-8}$) (Codd et al. 2013). Five of the loci contained genes (TERC, TERT, NAF1, OBFC1, RTEL1) with known involvement in telomere biology (Martinez and Blasco 2015, Schmidt and Cech 2015). A genetic risk score analysis combining lead variants at all seven loci in over 22,233 coronary artery disease cases and 64,762 controls showed an association of the alleles associated with shorter LTL with increased risk of CAD (21% (95% CI: 5–35%) per standard deviation in LTL, p = 0.014). Prior studies using Southern blotting to measure LTL have demonstrated that mean LTL attrition rate is about 30 bp per year (Vasa-Nicotera et al. 2005, Fitzpatrick et al. 2007, Baerlocher and Lansdorp 2003, Notaro et al. 1997, Rufer et al. 1999, Spyridopoulos et al. 2008, Spyridopoulos et al. 2009). The authors suggested that the per allele effect of the different SNPs on LTL in base pairs ranges from ~ 57 to 117 bp.

Mendelian Randomization Studies

A useful instrument for investigating causality between CAD and TL and thereby avoiding confounding is the use of genetic variants that result in shorter telomeres. The most obvious variants would be loss of function point mutations in telomerase (TERT and TERC) which are not associated with any known risk factors for CAD (confounders). If short telomeres would cause CAD, then the genetic variants are expected to result in higher odds-ratios for coronary heart disease. This approach is called Mendelian randomisation and has been successfully used, for example for cholesterol and C-reactive protein, to suggest causality with coronary heart disease (Palmer et al. 2013, Kamstrup et al. 2009, Burgess et al. 2012). Ideally, chosing candidate genes on different chromosomes additionally reduces the likelyhood that the genetic effect is due to linkage disequilibrium, where other genes in proximity have an implication for TL. There is currently no published Mendelian randomisation study investigating a link between TL and CAD.

Telomere Length in Progeroid Syndromes

Premature aging syndromes, such as Hutchinson–Gilford progeria syndrome (HGPS) or Werner syndrome demonstrate a cardiovascular phenotype resulting in severely reduced life span (Ghosh and Zhou 2014). The underlying molecular defect for both syndromes are transcriptional alterations and impaired DNA repair, with the latter leading to short telomeres and cellular senescence (Prokocimer et al.

2013). In HGPS, the permanently farnesylated state of progerin produces disruption of the nuclear lamina and a multitude of nuclear defects. T-loops can also form at interstitial telomeric sequences in a TRF2-dependent manner, forming an interstitial t-loop (ITL) (Simonet et al. 2011). Importantly, the association of TRF2 with interstitial telomeric sequences is stabilized by co-localization with A-type lamins (lamin A/C) (Wood et al. 2014). Therefore, mutations of lamin in HGPS that lead to progerin generation, and hence a reduction in levels of lamin A/C cause reduced ITL formation, and it is telomere loss, rather than telomere shortening, which results in cellular senescence. Of note, it is conceivable that telomere shortening in HGPS cells is the result of a combination of some telomeres of normal length with complete telomere loss in others, resulting in an average 'shorter' length per cell (metaphase cells contain 92 telomere foci). In HGPS, senescence of vascular smooth muscle cells in particular, but also accelerated oxidative stress, DNA damage, inhibition of telomerase, and marked telomere shortening, are all potential contributors to the generation of atherosclerotic plaque (Cao et al. 2011). In other genetic disorders associated with telomere shortening, heart failure is a common phenotype, such as in dyskeratosis congenita (DKC) (Chiang et al. 2004, Goldman et al. 2005). It is reasonabe to surmise that short telomeres alone are not necessarily causing premature atherosclerosis (DKC), but that they could potentially lead to accelerated senescence and therefore to accelerated atherosclerosis.

Mechanistic Explanations on How Short Telomere Length could Contribute to Coronary Heart Disease

There are several potential explanations of how short telomeres could lead to either accelerated or premature atherosclerosis. As outlined above, it seems unlikely from data in mouse models or from human diseases like DKC, that short telomere length itself can propagate coronary heart disease. The only dramatic onset of CAD in early adolescence paralleled by short telomeres is seen in HGPS, and here the main precipitating factor is DNA damage resulting in premature cellular senescence. We have also found (Fig. 5) that among myeloid and lymphoid populations respectively, telomere shortening in CAD appears to occur synchronously, which suggests either a predetermined shorter baseline in CAD, or shortening on stem cell level. The only exception in this pattern are cytotoxic CD8 lymphocytes, which seem to be affected by chronic inflammation and cytomegalovirus-infection, leading to accelerated telomere shortening in CAD. Therefore it is conceivable that shortened telomeres could propagate or accelerate atherosclerosis by the following mechanisms:

a) **Replicative senescence** occurs once short telomeres generate DNA damage signals in order to abort tumor generation. This can also lead to diminished regenerative capacity (Sahin and Depinho 2010, Jaskelioff et al. 2010), potentially reducing the self-healing ability of vascular endothelium in response to damage.

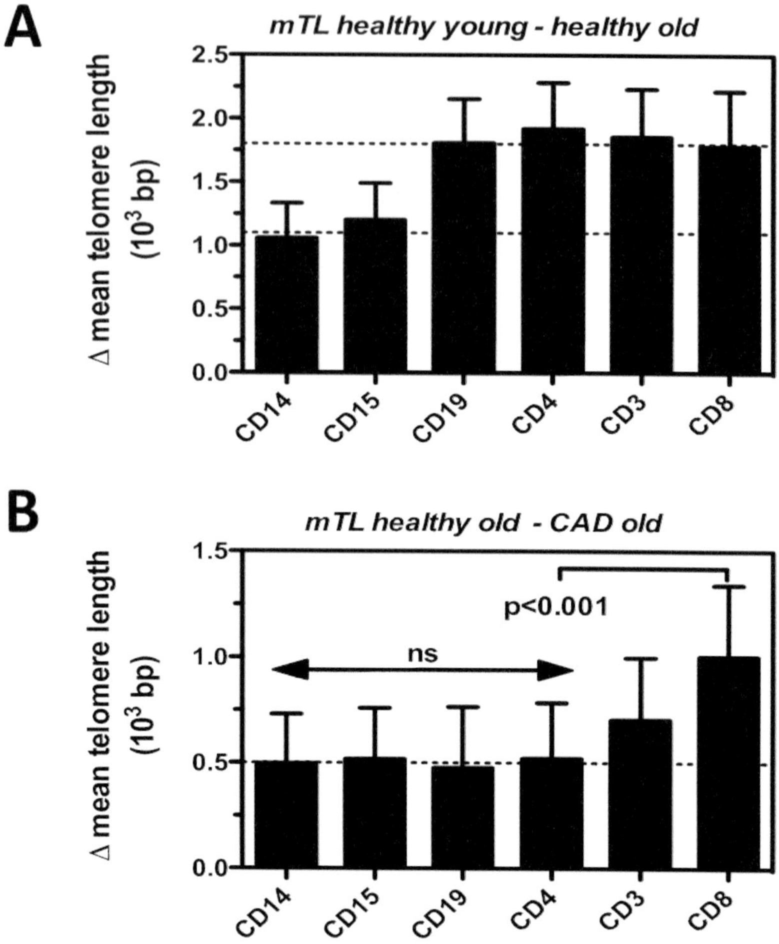

Fig. 5. Telomere length differences between different leukocyte subpopulations. A. Telomere length differences (mean ± SD) between young (n = 14) and old (n = 13) healthy volunteers is shown on the first graph. Mean age of the young population was 25 years, and mean age of the "old" population 65 years. We analysed telomere length in the 5 most common white blood cell populations: monocytes (CD14), granulocytes (CD15), B-lymphocytes (CD19), T-helper cells (CD4) and cytotoxic T-cells (CD8). The whole T-cell population (CD3) includes all CD4 and CD8 T cells. Telomere length differences in myeloid cells (CD14/CD15) were about 1100 base pairs and in lymphocytes 1800 bp, which would translate into an attrition rate of 28 bp/year and 45 bp/year respectively. B. Telomere length differences between CHD patients (n = 25) and age-matched healthy "old" volunteers (n = 13) was about 500 base pairs and almost identical in all myeloid and lymphoid subpopulations, which are exposed to (oxidative) stress in very different compartments of the organism, with one exception being the CD8 T-cells. Telomere length in CD8 cells seems to be uniquely regulated in CAD (Spyridopoulos et al. 2009).

b) **Telomere Position Effects** (TPE) can change gene expression at intermediate telomere lengths in cultured human cells. One study in humans demonstrated expression of *DUX4* 60 kB distant from the telomeric repeats, inversely correlating to telomere length (ranging between 7 and 20 kB) on one chromosome (Stadler et al. 2013). Once critical telomere length was reached, this lead to clinical symptoms in facioscapulohumeral muscular dystrophy (FSHD).

c) **Chromatin conformation changes** (chromosome looping) can bring the telomere close to genes up to 10 Mb away from the telomere when telomeres are long. When telomeres are short, the same loci can become separated. A recent study has revealed that many loci, including noncoding RNAs, may be regulated by telomere length (Robin et al. 2014). The authors have termed this process TPE-OLD for "**telomere position effect over long distances**".

d) **Telomere loss and telomere dysfunction** results in genomic instability, activation of DNA damage responses (DDR), mitochondrial dysfunction, and stem cell exhaustion, all of which have been implicated in the aging process (Sahin and Depinho 2010). HGPS is a good example of how this can result in premature atherosclerosis.

Conclusions

Despite the now well documented fact that CAD is associated with short telomeres, and how conversely short telomere length predisposes to CAD, there has been no proof of causality until now. Evidence from progeroid syndromes suggests that severe disruption of DNA repair results in shorter mean telomere length and cellular senescence, which in turn can cause premature atherosclerosis. Short telomere length in peripheral blood leukocytes itself can (1) have an inherited component without the co-existence of DNA damage, (2) indicate increased cell turnover, e.g., under systemic low grade inflammation, or (3) be a surrogate for accelerated DNA damage and cell senescence. In order to make more informed conclusions about the causal relationship between telomere length and CAD, further research is needed, both in clinical studies and on a molecular level. Finally, we might be able to learn from progeroid syndromes how DNA damage and telomere length are linked and thereby promote accelerated atherosclerosis.

Keywords: Telomeres, telomere length, coronary artery disease, progeroid syndromes

References

Anand, I.S., R. Latini, V.G. Florea, M.A. Kuskowski, T. Rector, S. Masson et al. 2005. C-reactive protein in heart failure: prognostic value and the effect of valsartan. Circulation. 112: 1428–1434.

Baerlocher, G.M. and P.M. Lansdorp. 2003. Telomere length measurements in leukocyte subsets by automated multicolor flow-FISH. Cytometry. 55A: 1–6.

Baerlocher, G.M., A. Roth and P.M. Lansdorp. 2003. Telomeres in hematopoietic stem cells. Ann. N. Y. Acad. Sci. 996: 44–48.

Baerlocher, G.M., I. Vulto, G. de Jong and P.M. Lansdorp. 2006. Flow cytometry and FISH to measure the average length of telomeres (flow FISH). Nat. Protoc. 1: 2365–2376.

Blackburn, E.H. 1991. Structure and function of telomeres. Nature. 350: 569–573.
Blasco, M.A. 2005. Telomeres and human disease: ageing, cancer and beyond. Nat. Rev. Genet. 6: 611–622.
Broer, L., V. Codd, D.R. Nyholt, J. Deelen, M. Mangino, G. Willemsen et al. 2013. Meta-analysis of telomere length in 19,713 subjects reveals high heritability, stronger maternal inheritance and a paternal age effect. Eur. J. Hum. Genet. 21: 1163–1168.
Brouilette, S., R.K. Singh, J.R. Thompson, A.H. Goodall and N.J. Samani. 2003. White cell telomere length and risk of premature myocardial infarction. Arterioscler. Thromb. Vasc. Biol. 23: 842–846.
Brouilette, S.W., J.S. Moore, A.D. McMahon, J.R. Thompson, I. Ford, J. Shepherd et al. 2007. Telomere length, risk of coronary heart disease, and statin treatment in the West of Scotland Primary Prevention Study: a nested case-control study. Lancet. 369: 107–114.
Burgess, S., A. Butterworth, A. Malarstig and S.G. Thompson. 2012. Use of Mendelian randomisation to assess potential benefit of clinical intervention. BMJ. 345: e7325.
Cao, K., C.D. Blair, D.A. Faddah, J.E. Kieckhaefer, M. Olive, M.R. Erdos et al. 2011. Progerin and telomere dysfunction collaborate to trigger cellular senescence in normal human fibroblasts. J. Clin. Invest. 121: 2833–2844.
Cawthon, R.M. 2002. Telomere measurement by quantitative PCR. Nucleic Acids Res. 30: e47.
Cawthon, R.M. 2009. Telomere length measurement by a novel monochrome multiplex quantitative PCR method. Nucleic Acids Res. 37: e21.
Chiang, Y.J., M.T. Hemann, K.S. Hathcock, L. Tessarollo, L. Feigenbaum, W.C. Hahn et al. 2004. Expression of telomerase RNA template, but not telomerase reverse transcriptase, is limiting for telomere length maintenance *in vivo*. Mol. Cell. Biol. 24: 7024–7031.
Chiu, C.P., W. Dragowska, N.W. Kim, H. Vaziri, J. Yui, T.E. Thomas et al. 1996. Differential expression of telomerase activity in hematopoietic progenitors from adult human bone marrow. Stem Cells. 14: 239–248.
Codd, V., C.P. Nelson, E. Albrecht, M. Mangino, J. Deelen, J.L. Buxton et al. 2013. Identification of seven loci affecting mean telomere length and their association with disease. Nat. Genet. 45: 422–427.
Epel, E.S., E.H. Blackburn, J. Lin, F.S. Dhabhar, N.E. Adler, J.D. Morrow et al. 2004. Accelerated telomere shortening in response to life stress. Proc. Natl. Acad. Sci. USA. 101: 17312–17315.
Farzaneh-Far, R., R.M. Cawthon, B. Na, W.S. Browner, N.B. Schiller and M.A. Whooley. 2008. Prognostic value of leukocyte telomere length in patients with stable coronary artery disease: data from the heart and soul study. Arterioscler. Thromb. Vasc. Biol. 28: 1379–1384.
Farzaneh-Far, R., J. Lin, E. Epel, K. Lapham, E. Blackburn and M.A. Whooley. 2010. Telomere length trajectory and its determinants in persons with coronary artery disease: longitudinal findings from the heart and soul study. PLoS One. 5: e8612.
Fitzpatrick, A.L., R.A. Kronmal, J.P. Gardner, B.M. Psaty, N.S. Jenny, R.P. Tracy et al. 2007. Leukocyte telomere length and cardiovascular disease in the cardiovascular health study. Am. J. Epidemiol. 165: 14–21.
Ghosh, S. and Z. Zhou. 2014. Genetics of aging, progeria and lamin disorders. Curr. Opin. Genet. Dev. 26: 41–46.
Goldman, F., R. Bouarich, S. Kulkarni, S. Freeman, H.Y. Du, L. Harrington et al. 2005. The effect of TERC haploinsufficiency on the inheritance of telomere length. Proc. Natl. Acad. Sci. USA. 102: 17119–17124.
Harley, C.B., A.B. Futcher and C.W. Greider. 1990. Telomeres shorten during ageing of human fibroblasts. Nature. 345: 458–460.
Haycock, P.C., E.E. Heydon, S. Kaptoge, A.S. Butterworth, A. Thompson and P. Willeit. 2014. Leukocyte telomere length and risk of cardiovascular disease: systematic review and meta-analysis. BMJ. 349: g4227.
Jakob, S., P. Schroeder, M. Lukosz, N. Büchner, I. Spyridopoulos, J. Altschmied et al. 2008. Nuclear protein tyrosine phosphatase Shp-2 is one important negative regulator of nuclear export of telomerase reverse transcriptase. J. Biol. Chem. 283: 33155–33161.
Jaskelioff, M., F.L. Muller, J.H. Paik, E. Thomas, S. Jiang, A.C. Adams et al. 2010. Telomerase reactivation reverses tissue degeneration in aged telomerase-deficient mice. Nature. 469: 102–106.
Kamstrup, P.R., A. Tybjaerg-Hansen, R. Steffensen and B.G. Nordestgaard. 2009. Genetically elevated lipoprotein(a) and increased risk of myocardial infarction. JAMA. 301: 2331–2339.
Kimura, M., R.C. Stone, S.C. Hunt, J. Skurnick, X. Lu, X. Cao et al. 2010. Measurement of telomere length by the Southern blot analysis of terminal restriction fragment lengths. Nat. Protoc. 5: 1596–1607.

Koch, S., A. Larbi, E. Derhovanessian, D. Ozcelik, E. Naumova and G. Pawelec. 2008. Multiparameter flow cytometric analysis of CD4 and CD8 T cell subsets in young and old people. Immun. Ageing. 5: 6.

Landmesser, U., S. Spiekermann, S. Dikalov, H. Tatge, R. Wilke, C. Kohler et al. 2002. Vascular oxidative stress and endothelial dysfunction in patients with chronic heart failure: role of xanthine-oxidase and extracellular superoxide dismutase. Circulation. 106: 3073–3078.

Martin-Ruiz, C.M., D. Baird, L. Roger, P. Boukamp, D. Krunic, R. Cawthon et al. 2014. Reproducibility of telomere length assessment: an international collaborative study. Int. J. Epidemiol. 44: 1749–1754.

Martinez, P. and M.A. Blasco. 2015. Replicating through telomeres: a means to an end. Trends Biochem. Sci. 40: 504–515.

McElligott, R. and R.J. Wellinger. 1997. The terminal DNA structure of mammalian chromosomes. EMBO J. 16: 3705–3714.

Notaro, R., A. Cimmino, D. Tabarini, B. Rotoli and L. Luzzatto. 1997. *In vivo* telomere dynamics of human hematopoietic stem cells. Proc. Natl. Acad. Sci. USA. 94: 13782–13785.

Oikawa, S., S. Tada-Oikawa and S. Kawanishi. 2001. Site-specific DNA damage at the GGG sequence by UVA involves acceleration of telomere shortening. Biochemistry. 40: 4763–4768.

Palmer, T.M., B.G. Nordestgaard, M. Benn, A. Tybjærg-Hansen, G. Davey Smith, D.A. Lawlor et al. 2013. Association of plasma uric acid with ischaemic heart disease and blood pressure: mendelian randomisation analysis of two large cohorts. BMJ. 347: f4262.

Petersen, S., G. Saretzki and T. von Zglinicki. 1998. Preferential accumulation of single-stranded regions in telomeres of human fibroblasts. Exp. Cell. Res. 239: 152–160.

Poch, E., P. Carbonell, S. Franco, A. Diez-Juan, M.A. Blasco and V. Andres. 2004. Short telomeres protect from diet-induced atherosclerosis in apolipoprotein E-null mice. Faseb J. 18: 418–420.

Prokocimer, M., R. Barkan and Y. Gruenbaum. 2013. Hutchinson-Gilford progeria syndrome through the lens of transcription. Aging Cell. 12: 533–543.

Robin, J.D., A.T. Ludlow, K. Batten, F. Magdinier, G. Stadler, K.R. Wagner et al. 2014. Telomere position effect: regulation of gene expression with progressive telomere shortening over long distances. Genes Dev. 28: 2464–2476.

Rufer, N., W. Dragowska, G. Thornbury, E. Roosnek and P.M. Lansdorp. 1998. Telomere length dynamics in human lymphocyte subpopulations measured by flow cytometry. Nat. Biotechnol. 16: 743–747.

Rufer, N., T.H. Brümmendorf, S. Kolvraa, C. Bischoff, K. Christensen, L. Wadsworth et al. 1999. Telomere fluorescence measurements in granulocytes and T lymphocyte subsets point to a high turnover of hematopoietic stem cells and memory T cells in early childhood. J. Exp. Med. 190: 157–167.

Sahin, E. and R.A. Depinho. 2010. Linking functional decline of telomeres, mitochondria and stem cells during ageing. Nature. 464: 520–528.

Samani, N.J., R. Boultby, R. Butler, J.R. Thompson and A.H. Goodall. 2001. Telomere shortening in atherosclerosis. Lancet. 358: 472–473.

Saretzki, G., M.P. Murphy and T. von Zglinicki. 2003. MitoQ counteracts telomere shortening and elongates lifespan of fibroblasts under mild oxidative stress. Aging Cell. 2: 141–143.

Schmidt, J.C. and T.R. Cech. 2015. Human telomerase: biogenesis, trafficking, recruitment, and activation. Genes Dev. 29: 1095–1105.

Simonet, T., L.E. Zaragosi, C. Philippe, K. Lebrigand, C. Schouteden, A. Augereau et al. 2011. The human TTAGGG repeat factors 1 and 2 bind to a subset of interstitial telomeric sequences and satellite repeats. Cell Res. 21: 1028–1038.

Southern, E.M. 1975. Detection of specific sequences among DNA fragments separated by gel electrophoresis. J. Mol. Biol. 98: 503–517.

Spyridopoulos, I., Y. Erben, T.H. Brummendorf, J. Haendeler, K. Dietz, F. Seeger et al. 2008. Telomere gap between granulocytes and lymphocytes is a determinant for hematopoetic progenitor cell impairment in patients with previous myocardial infarction. Arterioscler. Thromb. Vasc. Biol. 28: 968–974.

Spyridopoulos, I., J. Hoffmann, A. Aicher, T.H. Brümmendorf, H.W. Doerr, A.M. Zeiher et al. 2009. Accelerated telomere shortening in leukocyte subpopulations of patients with coronary heart disease: role of cytomegalovirus seropositivity. Circulation. 120: 1364–1372.

Stadler, G., F. Rahimov, O.D. King, J.C. Chen, J.D. Robin, K.R. Wagner et al. 2013. Telomere position effect regulates DUX4 in human facioscapulohumeral muscular dystrophy. Nat. Struct. Mol. Biol. 20: 671–678.

Valdes, A.M., T. Andrew, J.P. Gardner, M. Kimura, E. Oelsner, L.F. Cherkas et al. 2005. Obesity, cigarette smoking, and telomere length in women. Lancet. 366: 662–664.
van der Harst, P., G. van der Steege, R.A. de Boer, A.A. Voors, A.S. Hall, M.J. Mulder et al. 2007. Telomere length of circulating leukocytes is decreased in patients with chronic heart failure. J. Am. Coll. Cardiol. 49: 1459–1464.
Vasa-Nicotera, M., S. Brouilette, M. Mangino, J.R. Thompson, P. Braund, J.R. Clemitson et al. 2005. Mapping of a major locus that determines telomere length in humans. Am. J. Hum. Genet. 76: 147–151.
Vega, L.R., M.K. Mateyak and V.A. Zakian. 2003. Getting to the end: telomerase access in yeast and humans. Nat. Rev. Mol. Cell. Biol. 4: 948–959.
von Zglinicki, T., R. Pilger and N. Sitte. 2000. Accumulation of single-strand breaks is the major cause of telomere shortening in human fibroblasts. Free Radic. Biol. Med. 28: 64–74.
von Zglinicki, T. 2002. Oxidative stress shortens telomeres. Trends Biochem. Sci. 27: 339–344.
Wang, J.C., J.K. Warner, N. Erdmann, P.M. Lansdorp, L. Harrington and J.E. Dick. 2005. Dissociation of telomerase activity and telomere length maintenance in primitive human hematopoietic cells. Proc. Natl. Acad. Sci. USA. 102: 14398–14403.
Wood, A.M., J.M. Rendtlew Danielsen, C.A. Lucas, E.L. Rice, D. Scalzo, T. Shimi et al. 2014. TRF2 and lamin A/C interact to facilitate the functional organization of chromosome ends. Nat. Commun. 5: 5467.
Wright, W.E., M.A. Piatyszek, W.E. Rainey, W. Byrd and J.W. Shay. 1996. Telomerase activity in human germline and embryonic tissues and cells. Dev. Genet. 18: 173–179.
Wright, W.E., V.M. Tesmer, K.E. Huffman, S.D. Levene and J.W. Shay. 1997. Normal human chromosomes have long G-rich telomeric overhangs at one end. Genes Dev. 11: 2801–2809.

Chapter 6

Environmental and Occupational Exposure and Telomere Length

Natalia Pawlas

INTRODUCTION

There is an increased exposure to xenobiotics including new compounds and nanoparticles in human environment. It is a hazard of progressive industrialization. Long-term, low-level exposure to multiple chemicals is a characteristic feature of modern, highly developed societies. According to the reports published by the Centers for Disease Control and Prevention (CDC 2009), this was reflected by an increase in the number of chemicals that can be found in human body (blood or urine) at detectable concentrations even in the case of solely environmental exposure. The Fourth National Report on Human Exposure to Environmental Chemicals (CDC 2015, updated February 2015) stresses the presence of more than 150 chemicals, previously shown in occupational exposure, in biological samples from environmentally exposed population. While some of those chemicals are well studied, the effects of long-term exposure and mechanism of action of others are unknown. It shows that in general population there are detectable concentrations of carcinogens (acrylamide), endocrine disruptors, bisphenol A, pesticides, metals (including arsenic, cadmium, lead, mercury), perfluorochemicals, phthalates, phytoestrogens, organochlorine substances (including dioxin-like), polycyclic aromatic hydrocarbons and others. Many of these compounds are also contaminants at work places as well, usually in higher concentrations. Persistent and bio-accumulative compounds, endocrine disruptors and heavy metals used in modern industry as well as wastes containing phthalates, perfluorinated chemicals, are particularly dangerous since they may affect human reproduction, nervous system development, lead to obesity

Department of Pharmacology, School of Medicine with the Division of Dentistry in Zabrze, Medical University of Silesia, Jordana 38 str., 41-808 Zabrze, Poland.
Email: n-pawlas@wp.pl

and cancer (EEA Report 2010), and result in susceptibility or development of other non-communicable diseases. Some mechanisms of action are unknown and not fully understood, and some new studies show results which in turn lead to change in the International Agency for Research on Cancer (IARC) classification from one group to another, as it was recently done, e.g., for inorganic lead (change from group 2B to 2A). Recent data shows that low dose chemical exposure in modern society may lead to genome instability (e.g., single or double stranded DNA break, impairment of processes of DNA repair, mutations, structural or numerical chromosomal changes, epigenetic modifications) which, further, may lead to development of cancer (Langie et al. 2015). On the other hand, we observe an increase in morbidity of some diseases as well as their earlier onset.

According to European Environment Agency (EEA 2013) report, the degradation of environment may contribute to substantial increase in morbidity of obesity, diabetes, cardiovascular diseases, nervous system diseases and cancer, which are major public health concerns in Europe. This degradation of natural environment is caused, for example, by air pollution (particulate matter), noise, chemicals, poor quality of water, pesticides in the environment and environmental tobacco smoke (EEA 2013). The WHO estimates the environmental burden of disease in Europe at 15–20% of total death and 14–20% of disability-adjusted life years (Pruss-Ustun and Corvalan 2006). Human exposure occurs *via* absorption from respiratory and gastrointestinal tract by toxicants in water, air, food, indoor dust and consumer products.

Telomeres, Diseases and Exposure

During recent years, telomere length has been gaining the interest of researchers in the context of various pathological conditions (diabetes mellitus, metabolic syndrome, cardiovascular diseases, neoplasms), as well as due to its potential association with occupational and environmental exposure to xenobiotics (heavy metals, polycyclic aromatic hydrocarbons, particle pollution, etc.). Since shortening of the telomeres is age specific, it is considered a biomarker of cellular ageing. In previous cross-sectional and some prospective studies, telomere shortening turned out to be associated with arterial hypertension, atherosclerosis, ischemic heart disease, type 2 diabetes mellitus and (albeit not universally) neoplastic diseases (Wentzensen et al. 2011, Willeit et al. 2010a, 2010b, Willeit et al. 2014).One study in Normative Aging Study Cohort shows that telomere shortening was more pronounced in patients who would develop cancer several years before the onset of the disease. In the period preceding cancer diagnosis the shortening was decelerated (Hou et al. 2015). Although prospective epidemiological studies dealing with the problem in question are sparse, there is some evidence suggesting that telomere shortening may be associated with increased risk of metabolic syndrome, as well as with an unfavourable prognosis in bladder neoplasms and idiopathic pulmonary fibrosis (Revesz et al. 2014, Russo et al. 2014, Stuart et al. 2014). All this evidence points to another potential harmful effect of exposure to xenobiotics, as it was also shown to result in telomere shortening. Previous studies demonstrated that individuals exposed to xenobiotics, such as polycyclic aromatic hydrocarbons, N-nitrosamines, pesticides and cadmium, presented with shorter telomeres than control ones, while some others were associated with longer telomeres (Zhang et al. 2013). Also, psychosocial stress was proven to shorter telomeres (Savolainen et al. 2014, Starkweather et al. 2014).

The developed methods of assessing telomere length using quantitative polymerase chain reaction (qPCR) (Cawthon 2002) are throughput and allow examining large populations. On the other hand, that widely used method measures relative telomere length and most of the studies refer to such measurements. The established method of absolute telomere length measurements is less frequently used (O'Callaghan and Fenech 2011). In the relative telomere length measurements, one must remember that those are the results of measurements which are related to a reference DNA (which is quite frequently a mixture of pooled DNA samples from control or unexposed group). That is why the interpretation of results may be difficult since the fact that exposed group has longer telomeres than control group may mean that either there is some mechanism which leads to telomere elongation or there are mechanisms impairing the physiological telomere shortening in the exposed group, and their telomeres shortened more slowly than in unexposed ones. Both mechanisms, leading to faster telomere shortening in exposed ones and to slower in control ones, or to telomere elongation are not fully understood and unknown. Some authors suggest that premature telomere shortening in occupationally exposed subjects may be related to increased oxidative stress, which occurs during exposure to N-nitrosamines (Li et al. 2011), heavy metals (Pawlas et al. 2015a, 2015b, Wu et al. 2012), or tobacco smoke (Babizhayev and Yegorov 2011). Wong et al. (2014c) showed that global DNA hypomethylation may play a role in accelerated telomere shortening in boilermakers (Wong et al. 2014c). In case of factors leading to longer telomeres in exposed ones, the potential explanation is related to increased telomerase activity as in arsenic (Li et al. 2012, Ferrario et al. 2009), dioxins (Sarkar et al. 2006) and benzene exposure (Bassig et al. 2014). However, other studies showed no changes in telomerase expression and methylation in subjects with telomere elongation (Dioni et al. 2011).

Aging of population of developed countries, including workers, is the other problem nowadays. Aging of workers contributes to their health, diseases, disabilities and abilities to work. Establishing determinants of healthy aging is important to prevent early onset of disabilities. Long-term exposure to low-level environmental and low- or high-level occupational stressors may lead to earlier onset and higher prevalence of some diseases. That is why the possible nutritional or other life-style intervention which may prevent accelerated telomere shortening would be promising (Cassidy et al. 2010, Garcia-Calzon et al. 2015, Sun et al. 2012).

Here we summarize the studies on environmental and occupational stressors and telomere length (Tables 1–3).

Toxic metals

Metals are elements characterized by similar specific physicochemical properties. This is a heterogeneous group of elements. Some of the metals like sodium, potassium or calcium are common in human body; others like selenium or iron are recognized as trace elements. Other metals may be toxic for living organisms. Toxic metals

Table 1. Summary of the studies on effects of chemicals or mixtures on telomere length in humans.

Chemical name/Mixture	Type of study and exposure	Participants	Effects on telomere length	Reference
Lead	Cross-sectional, occupational exposure	144 battery plant workers, aged 18–55	Negative correlation with lead body burden, weak positive with blood lead level.	(Wu et al. 2012)
Lead	Cross-sectional with control group, Occupational exposure	334 lead and zinc smelters, aged 19–62 and 60 non-exposed ones, aged 25–57	Telomere length was shorter in lead exposed and there was a dose-dependent effect. Relative telomere length was negatively associated with oxidative stress.	(Pawlas et al. 2015b)
Lead	Cross-sectional, environmental exposure	99 children, eight-years of age	There was a significant negative correlation between blood lead level and telomere length. Also, prenatal tobacco exposure was a significant factor leading to telomere shortening.	(Pawlas et al. 2015a)
Lead	Cross-sectional, environmental exposure	6,796 participants, aged 20–84	There was no association between blood lead level and telomere length.	(Zota et al. 2015)
Lead	Cross-sectional with control group, environmental exposure	227 living in the vicinity of electronic waste recycling workshops and 93 control ones, adults	There was no association between placental lead level and telomere length.	(Lin et al. 2013)
Cadmium	Cross-sectional, environmental exposure	6,796 participants, aged 20–84	Blood and urine cadmium levels were negatively associated with telomere length.	(Zota et al. 2015)
Cadmium	Cross-sectional with control group, environmental exposure	227 living in the vicinity of electronic waste recycling workshops and 93 control ones, adults	Placental cadmium level was negatively associated with placental telomere length.	(Lin et al. 2013)
Cadmium	Cross-sectional with control group, Occupational exposure	334 lead and zinc smelters, aged 19–62 and 60 non-exposed ones, aged 25–57	There was no effect of cadmium exposure on telomere length.	(Pawlas et al. 2015b)
Cadmium	Cross-sectional, environmental exposure	99 children, eight-years of age	There was no effect of cadmium exposure on telomere length.	(Pawlas et al. 2015a)

Table 1 contd. ...

Fractional Order PID and ADR Controls

This book explores the design and analysis of fractional-order and active disturbance rejection control, examining both the theoretical foundations and their practical applications.

It covers fractional-order proportional-integral-derivative (PID) control, fractional-order active disturbance rejection (ADR) control, and the combined fractional-order PID-ADR control. The book begins with an analysis of the three-parameter fractional-order PID controller, demonstrating its application to the permanent magnet synchronous motor (PMSM) speed servo system, due to its comprehensive inclusion of proportional, integral, and differential elements. It then delves into active disturbance rejection control and periodic disturbance compensation, comparing the performance of each controller based on various parameters. This comparison enables readers to critically evaluate the advantages and limitations of each approach before implementation. Offering a thorough guide to fractional-order and active disturbance rejection control, the book also includes numerical methods for assessing and developing these systems.

The book will be of particular interest to professionals working with numerical methods, fractional-order systems and control, PID controller, active disturbance rejection, control design, and production and is especially relevant to those in mechanical, industrial, and electrical engineering.

...Table 1 contd.

Chemical name/Mixture	Type of study and exposure	Participants	Effects on telomere length	Reference
Arsenic	Cross-sectional with control group, environmental exposure	161 highly-exposed women and 41 controls, adults	Arsenic exposure through drinking water was associated with longer telomeres.	(Li et al. 2012)
Arsenic	Cross-sectional study, population recruited out of two prospective studies	1,799 adults aged 43.5 ± 10.6 from one study and 167, aged 35.5 ± 9.7 from the second one	Telomere length was tested in 167 samples and was positively associated with arsenic exposure.	(Gao et al. 2015)
Zinc	Cross-sectional study	155 children aged three, 155 aged six and 154 aged nine years	Zinc plasma level was inversely associated with telomere length in children.	(Milne et al. 2015)
Selenium	Cross-sectional, occupational and environmental exposure (two cohorts)	99 children, eight-years of age 334 lead and zinc smelters, aged 19–62 and 60 non-exposed ones, aged 25–57	There were no correlations, both in children and in adults, between selenium levels in blood or serum and telomere length.	(Pawlas et al. 2015a, 2015b)
Pesticides	Cross-sectional with control group, occupational exposure, Absolute telomere length	124 exposed, aged 17–78 and 62 non-exposed aged 23–76	Exposed subjects had significantly shorter telomeres than non-exposed ones. Questionnaire on retrospective use of pesticides: glyphosate (92%), flumetralin 28,5% and other (carbofuran, clomazone, imidacloprid, sulfentrazone, ipridione, deltamethrin, paraquat, butralin, promocarb hydrochloride, mancozeb, 2,4-dimethylamine).	(Kahl et al. 2016)
Pesticides	Prospective cohort, occupational exposure	1,234 participants, aged 40 and over	Commercial pesticide applicators had shorter telomere length than private applicators. Duration of lifetime exposure to pesticides associated with shorter telomere length alachlor, metolachlor, trifluralin, 2,4-dichlorophenoxyacetic acid (2,4-D), permethrin, toxaphene.	(Hou et al. 2013)

Table 1 contd. ...

...Table 1 contd.

Chemical name/Mixture	Type of study and exposure	Participants	Effects on telomere length	Reference
Pesticides	Prospective cohort, occupational exposure	568 males, aged 31–94	2,4-D, diazinon, butylate—duration of lifetime exposure was associated with shorter telomere length. Recent use of malathion was associated with shorter telomeres, while recent use of alachlor with longer telomeres.	(Andreotti et al. 2015)
Persistent organic pollutants (POPs), polychlorinated biphenyls (PCBs)	Cross-sectional study, NHANES cohort	4,260 participants, aged 20 and over	POPs with high toxic equivalency factor and high affinity to bind to aryl hydrocarbon receptor (PCB-126 and PCB-169) were associated with longer telomeres.	(Mitro et al. 2015)
Persistent organic pollutants (POPs), polychlorinated biphenyls (PCBs)	Cross-sectional	84 participants, aged 40 and over	Most organochlorine pesticides and polychlorinated biphenyls were associated with longer telomere length. The strongest association were found with p,p'-dichlorodiphenyldichloroethylene, PCB99, PCB153, PCB180, PCB183 and PCB187. The association between telomere length and the POPs concentration was non-linear and more pronounced in lower level of exposure.	(Shin et al. 2010)
Persistent organic pollutants (POPs), polychlorinated biphenyls (PCBs)	Cross-sectional study, NHANES cohort	2,413 adults, geometric mean of age 42.8	Sum of PCBs was associated with longer telomeres in environmental exposure. Additionally, 1,2,3,4,6,7,8-heptachlorodibenzofuran and 1,2,3,6,7,8-hexachlorodibenzo-p-dioxin were associated with longer telomeres.	(Scinicariello and Buser 2015)
Benzene	Cross-sectional with control group, occupational exposure	22 high-exposed aged 36 ± 8 years, 21 low-exposed aged 35 ± 8 years and 43 controls aged 36 ± 7 years	High levels of occupational exposure to benzene were associated with longer telomeres.	(Bassig et al. 2014)

Table 1 contd. ...

...Table 1 contd.

Chemical name/Mixture	Type of study and exposure	Participants	Effects on telomere length	Reference
N-nitrosamines	Cross-sectional, occupational exposure	157 workers from rubber industry, adults	Occupational exposure to N-nitrosamines in rubber industry was associated with shorter telomeres; there was also inverse association between telomere length and p-toluidine.	(Li et al. 2011)
Coal tar pitch	Cross-sectional study with control group, occupational exposure	180 exposed and 145 control ones	Exposure to coal tar pitch was associated with shorter telomere length.	(Wang et al. 2015)
Tobacco smoke	Cross-sectional study, active smokers	147 smokers, aged 25–65	Tobacco exposure (lifetime accumulated) was associated with a damage to telomeres (shorter telomeres).	(Verde et al. 2015)
Tobacco smoke	Cross-sectional study, female twins, Southern blot method	1,122 women, aged 18–72	There was a dose-dependent telomere shortening. Each pack-year was related to additional loss of telomere length by 5 base pairs.	(Valdes et al. 2005)
Tobacco smoke	Two prospective cohorts—Nurses' Health Study and Health Professionals Follow-up Study	184 cases with bladder cancer (58 – never smokers) and 192 controls (60 never smokers), adults	They observed a significant reduction of telomere length in cases and control who smoked comparing to those who never smoked.	(McGrath et al. 2007)
Tobacco smoke	Cross-sectional study, prenatal exposure	104 children, aged 4–14 (9 ones exposed prenatally)	Telomeres of children whose mother smoked during pregnancy were significantly shorter.	(Theal et al. 2013)
Tobacco smoke	Prospective cohort, prenatal exposure	86 newborns of mother who never smoked (25), were passive smokers (31) and were active smokers (30)	They observed a dose-response reduction of telomere length in newborns that were prenatally exposed tobacco smoke (either through passive or active smoking of their mothers).	(Salihu et al. 2014)

Table 1 contd. ...

...Table 1 contd.

Chemical name/Mixture	Type of study and exposure	Participants	Effects on telomere length	Reference
Tobacco smoke	Cross-sectional study, prenatal exposure	99 children (10 exposed prenatally)	Children of mothers who smoked during pregnancy had by 13% significantly shorter telomeres compared to references whose mothers did not smoke.	(Pawlas et al. 2015a)
Hairdressers' exposure, (aromatic amines, oxidants)	Cross-sectional study with control group, occupational exposure	295 hairdressers and 92 controls, aged 18–55	Hairdressers had significantly shorter telomeres than control ones.	(Li et al. 2016)
Welding fumes	Prospective cohort, occupational exposure	48 boilermakers, adults	Telomere shortening was associated with recent PM2.5 exposure (one month). Longer cumulative exposure did not affect telomere length.	(Wong et al. 2014a)
Welding fumes	Cross-sectional study with control group, occupational exposure	101 welders, aged 23–60, and 127 controls, aged 23–56	Telomeres were significantly shorter with working years as a welder (model with several adjustments including age); however, there was no difference between exposed and non-exposed ones.	(Li et al. 2015)
Fine particulate matters PM2.5, Metal particles	Prospective cohort, occupational exposure	87 Welder and boiler repairing staff, aged 18 and over	Negative correlation with increased systemic inflammation, C-reactive protein and Serum Amyloid A. No association with cumulative PM exposure. Telomere length was positively associated with LINE-1 and Alu methylation. LINE-1 methylation modified the rate of telomeric change.	(Wong et al. 2014c, 2014b)
Particulate matters PM10 and PM1, steel workers	Prospective cohort, short-term (three days) occupational exposure	63 subjects, aged 27–55	There was a dose-dependent increase in telomere length in subjects exposed to PM during short exposure. There was no change in telomerase expression.	(Dioni et al. 2011)
Traffic pollution	Cross-sectional study with control group, occupational exposure	77 traffic officers and 57 office workers (control group), adults	Telomere length of traffic officers working in highly polluted area (high traffic intensity) was significantly reduced compared to office workers, and was negatively associated with measured levels of benzene and toluene.	(Hoxha et al. 2009)

Table 1 contd. ...

...Table 1 contd.

Chemical name/ Mixture	Type of study and exposure	Participants	Effects on telomere length	Reference
Air pollution/ traffic pollution	Cross-sectional study with control group, occupational exposure	120 truck drivers, 120 office workers	Short term exposure to particulate matter was associated with increased telomere length, while longer exposure led to telomere shortening.	(Hou et al. 2012)
Air pollution	Prospective study, Veteran Affairs Normative Aging Study cohort	165 never-smoking adult men	Long term exposure to air pollution, particularly particles of black carbon, was associated with telomere shortening.	(McCracken et al. 2010)
Indoor air pollution/ biomass smoke	Cross-sectional study	25 adult women	Women with high exposure had 43% shorter telomeres than women with low exposure.	(Shan et al. 2014)
Mixed exposure to chemicals and physical agents	Cross-sectional with control group, occupational exposure	120 workers of car mechanical workshops, 120 non-exposed controls, adults	Telomere length in exposed group was significantly lower.	(Eshkoor et al. 2012)

that require the most concern are as follows: lead, cadmium, mercury and arsenic. Exposure to them is common in polluted environment and in occupational settings and must be controlled.

Lead exposure was examined in a few studies. One *in vitro* study showed the telomere shortening in lead exposed hepatocytes (Liu et al. 2004). Another *in vitro* study suggested that lead ions could displace sodium and potassium ions from telomere G-quadruplexes, which in turn lead to their instability and damage during replication (Pottier et al. 2013).

In a study on occupational lead exposure, Wu et al. (2012) showed a negative association between telomere length and body lead burden and exposure duration in 144 adult battery workers. In that study, there was observed a weak positive association with blood lead level only in the very high exposed group, i.e., in workers whose blood lead level was higher than 40 μg/dl or urinary lead level was higher than 7 μg/dl (Wu et al. 2012). Pawlas et al. (2015b) showed that lead exposure lead to significantly shorter telomere length in 334 lead and zinc smelter and 60 non-exposed controls. That association was dose-dependent and increasing blood lead levels were associated with shorter telomere length. The authors also studied the oxidative stress (malonylodialdehyde level and total oxidative status) and suggested that telomere shortening in lead exposure may be related to oxidative stress (Pawlas et al. 2015b).

With respect to environmental exposure, studies are not unanimous; however, it is probably due to different level of exposures. Zota et al. (2015) carried out a study in a NHANES cohort of patients (NHANES, National Health and Nutrition

Table 2. Summary of the *in vitro* studies on effects of chemicals or mixtures on telomere length.

Chemical name	Type of study	Material	Effects on telomere length	Reference
Polychlorinated biphenyls PCBs	*In vitro* study	HaCaT Human keratinocyte cell line	Dose-dependent decrease of telomere length in cells exposed to 2-(4'-chlorophenyl)-1,4-benzoquinone.	(Jacobus et al. 2008)
Polychlorinated biphenyls PCBs	*In vitro* study	Human myeloblastic/ promyelocytic leukaemia cell line HL-60	Telomere shortening and telomerase downregulation, with upregulation of some shelterin complex.	(Xin et al. 2016)
Lead	*In vitro* study	Human EJ30 bladder cell carcinoma cell line—clone B3	Telomere loss, induction of histone γH2Ax foci.	(Pottier et al. 2013)
Lead	*In vitro* study	Hepatocytes L-02 cell line Hepatoma SMMC-7721 cell line	Telomere shortening in hepatocyte L-02 line. Extended telomeres in hepatoma SMMC-7721 line.	(Liu et al. 2004)
Selenium	*In vitro* study	Hepatocytes L-02 cell line Hepatoma SMMC-7721 cell line	Decelerated telomere loss and extended telomeres in hepatocyte L-02 line. No significant effect on hepatoma SMMC-7721 line.	(Liu et al. 2004)
Zinc	*In vitro* study	Hepatocytes L-02 cell line Hepatoma SMMC-7721 cell line	Helps to maintain telomere length in hepatocyte L-02 line. Shortening of telomeres in hepatoma SMMC-7721 line	(Liu et al. 2004)
Iron	*In vitro* study	Hepatocytes L-02 cell line Hepatoma SMMC-7721 cell line	Slight increase in telomere length in hepatocyte L-02 line. Extended telomeres in hepatoma SMMC-7721 line.	(Liu et al. 2004)
Chromium	*In vitro* study	Hepatocytes L-02 cell line Hepatoma SMMC-7721 cell line	Slight decrease in telomere length in hepatocyte L-02 line. Extended telomeres in hepatoma SMMC-7721 line.	(Liu et al. 2004)
Arsenic	*In vitro* study	Mononuclear cells isolated from male and female human umbilical cord (huUCB)	*In vitro* exposure to arsenic, at environmental concentrations, increased telomere length, but in higher concentration, decreased TL. Expression of telomerase reverse transcriptase (TERT) showed similar pattern.	(Ferrario et al. 2009)

Table 3. Summary of the studies on effects of psychosocial stress on telomere length in humans.

Factor name	Type of study and exposure	Participants	Effects on telomere length	Reference
Work-type/ employment status	Cross-sectional	981 subjects working full-time, half-time, physical and interpersonal stressors, aged 45–84	No effect	(Fujishiro et al. 2013)
Work-type/ employment status	Cross-sectional Sister Study	608 women working moderate or substantial work in past, currently full time working, non-employed, part-time working, adults	Women with moderate or substantial past work history (more than 20 years full time comparing to those working 1–5 years) or those currently working full-time had shorter telomeres.	(Parks et al. 2011)
Social disadvantage (harsh environment, poverty, mother's education, etc.)	Cross-sectional	40 African American boys, aged nine	Boys who were brought up in harsh environment had significantly shorter by 19% telomeres than their colleagues who were brought up in nurturing environment.	(Mitchell et al. 2014)
Childhood maltreatment	Cross-sectional	31 adults (22 women and nine men), aged 18–64	Those who reported childhood maltreatment (e.g., emotional abuse, emotional neglect, physical abuse, physical neglect, sexual abuse) had significantly shorter telomeres.	(Tyrka et al. 2010)
Childhood adverse events	Case-control study	321 participants with anxiety disorder and 653 controls, aged 30–87	There was no difference in telomere length between cases and controls. Shorter telomere length was associated with a greater number of childhood adversities.	(Kananen et al. 2010)
Social deprivation	Randomized-controlled trial of Foster care placement	136 children, aged 6–30 months, who were residents of institutional care or foster parents care	Children with greater exposure to institutional care had shorter telomeres.	(Drury et al. 2012)
Early life stress and separation	Cross-sectional study	215 participants aged 63.7 ± 2.9 years who were separated from parents in their childhood and 1271 participants, aged 60.9 ± 2.7 who were not separated.	Isolated separation or reported traumatic experiences were not associated with telomere shortening. Participants who were separated in childhood and reported traumatic experiences had shorter telomeres than those who were not separated.	(Savolainen et al. 2014)

Examination Survey). They did not find any association between environmental exposure to lead and telomere length in adult populations, but the mean blood lead level in NHANES population was low (1.67 μg/dl). Likewise, environmental exposure to lead (measured in placental tissue) was not associated with telomere length (Lin et al. 2013). In Polish eight-years-old children population, Pawlas et al. (2015a) observed telomere shortening, but the mean blood lead level of this group was 3.28 μg/dl, which was higher than in the NHANES cohort (Pawlas et al. 2015a, Zota et al. 2015).

Contradictory results are obtained in cadmium exposed populations. Environmental exposure to cadmium was associated with shorter telomere length in placental tissue of pregnant women living in the vicinity of electronic waste recycling site (Lin et al. 2013). Zota et al. (2015) also showed decrease of telomere length in NHANES adult population with mean blood cadmium level 0.44 μg/l (Zota et al. 2015). In a study by Pawlas et al. (2015a), no effect on telomere length was shown in eight-years-old children with much lower level of cadmium in their blood (0.18 μg/l) (Pawlas et al. 2015a).

With respect to metal exposure, there have been consistent studies showing that environmental exposure to arsenic results in longer telomeres and increased telomerase expression, which may partially explain the carcinogenic effects of arsenic, which is a recognized IARC 1 group carcinogen (Li et al. 2012, Ferrario et al. 2009, Gao et al. 2015). Li et al. (2012) examined 161 women living in San Antonio de los Cobres in Argentina and 41 women from surrounding villages. It was shown that public drinking water in San Antonio de los Cobres contains much more arsenic than in the surrounding villages. They showed that urinary arsenic was positively associated with telomere length. Also, urinary arsenic was associated with telomerase reverse transcriptase (TERT) transcripts; however, they did not find relation between TERT expression and telomere length. Gao et al. (2015) studied telomere length in 167 samples and found that it was positively associated with arsenic exposure. They also showed that urinary arsenic was inversely associated with expression of genes that codes shelterin complex components and positively with genes involved in telomeric DNA repair (Gao et al. 2015). *In vitro* exposure of mononuclear cells isolated from male and female human umbilical cord (huUCB) to arsenic, at environmental concentrations, increased telomere length, but in higher concentration decreased telomere length. Expression of TERT showed similar pattern, however more pronounced increase at low level was observed in female cells (Ferrario et al. 2014).

Zinc is considered as a trace element in human body. In a study by Milne et al. (2015), zinc plasma level was inversely associated with telomere length in children. The mechanism is unknown. However, the authors suggested that labile zinc may be responsible for oxidative stress generation and thus, increase in telomere sequence deletions (Milne et al. 2015).

Tobacco smoke

Tobacco smoke is a mixture of multiple chemical compounds including nicotine, particulate matters, polycyclic aromatic hydrocarbons (PAHs), 4-aminobiphenyl,

2-naphtylamine, o-toluidine, resorcinol, cresols, naphthalene, carbon monoxide, nitrogen oxides, hydrogen cyanide, dienes, phenols, formaldehyde, acetaldehyde, acrolein, cadmium, lead, arsenic and others (Rustemeier et al. 2002). Some of them are established carcinogens—benzo[a]pyrene and others PAHs, N-nitrosamines, aromatic amines, aldehydes, benzene, cadmium and other compounds (Hecht 2002). Toxic is both main stream and side stream (also referred as second hand smoke or environmental tobacco exposure). Exposure results in oxidative stress, inflammatory response, sensitization and carcinogenesis (Pappas 2011). Besides other widely-known mechanism of action, smoke compounds form DNA adducts and affect cellular metabolism and exert epigenetic changes (Hecht and Samet 2007).

Tobacco smoke exposure, both in an active and in passive smoker, seems to decrease telomere length. Valdes et al. (2005) conducted a research in female twins. They recruited 45 pairs of monozygotic twins and 516 pairs of dizygotic twins. They conducted a cross-sectional study in a cohort recruited from TwinsUK Adult Twin Registry. They observed a dose dependent relation between telomere length loss and pack-years. The other interesting factor leading to telomere shortening in that study was obesity (Valdes et al. 2005). Verde et al. (2015) measured telomere length in 147 healthy smokers as well as their ability to metabolise nicotine. The cigarette smoking was confirmed using exhaled carbon monoxide method. They observed a significant association between reduction of telomere length with the lifetime accumulated smoking exposure and there was no association between telomere length and nicotine metabolite ratio (Verde et al. 2015). McGrath et al. (2007) studied the telomere length in men and women with bladder cancer, in control group, and did not find any association between both in cancer and in healthy group; however, they stated that they might not have enough power for those analyses in Nurses' Health Study Cohort. In the second studied cohort (Health Professionals Follow-up Study), they examined similar parameters in 123 patients with bladder cancer and in 125 controls, they found significant reduction in telomere length in control smokers and pack-years of smoking; however, it did not affect the risk of bladder cancer. When they combined two cohorts, the results were significant and they observed the significant reduction of mean telomere length in all categories (McGrath et al. 2007).

Also, prenatal exposure to tobacco smoke seems to be an important factor for telomere length in future postnatal life. Theal et al. (2013) studied telomere length in 104 children aged 4–14 years and they found significantly shorter telomeres in children whose mothers smoked during pregnancy; however, the data on smoking was obtained from the questionnaires (Theal et al. 2013). Similar results were obtained by Pawlas et al. (2015a). They showed a significant reduction of telomere length (by 13%) in eight-years-old children whose mother smoked during pregnancy (Pawlas et al. 2015a). Those results of previously mentioned studies, where exposure was assessed by questionnaires, are in agreement with another study, where exposure level was determined by measurements of cotinine concentrations in maternal saliva (Salihu et al. 2014). In the prospective study, they measured telomere length in newborns of 86 mothers, who previously were asked if they smoked; their levels of cotinine in the saliva were assessed as well. Finally, mothers were divided into three groups—non-smokers, passive smokers and active smokers. The measurements of telomere length in their children clearly showed dose-response relationship between maternal smoking exposure and telomere length reduction in children (Salihu et al. 2014).

Air pollution

Air pollution, both in environmental (indoor and outdoor) as well as occupational setting, is quite frequently assessed by exposure to particulate matter (PM). That term refers to particles suspended in air. Such particles, which constitute a heterogeneous group, can be coated with chemicals, like polycyclic aromatic hydrocarbons (PAHs), nitrogen oxides or can be small metal particles or simply black carbon. The concentrations of PM10 and PM2.5 are frequently measured since they represent the fraction of particulate matters whose aerodynamic diameters are as small as 10 or 2.5 μm, respectively, and they can easily be inhaled by human lungs. The concentrations of PM10 and PM2.5 represent the quality of air. They are emitted into air through natural processes like volcano eruptions and by anthropogenic processes, mainly industrial ones, but also by combustion of fuels in houses and in cars. Therefore, due to heterogeneity of the term "particulate matter", it is difficult to directly compare different studies.

Hou et al. (2012) studied the effect of air pollution, measured by ambient particulate matter (PM10 and PM2.5), on telomere length in the inhabitants and workers of Beijing. They recruited 120 truck drivers and 120 office workers, obtained the data on PM10 from the local monitoring stations, while PM2.5 was measured using personal monitors. It occurred that truck drivers had significantly longer telomeres than office workers. They showed that short term exposure to air pollution may increase telomere length, while longer exposure (14 days) led to decrease in telomere length. That finding may correspond to inflammatory response (Hou et al. 2012). Consistent results for short term exposure in steel workers exposed to particulate matter were obtained by Dioni et al. (2011), while the same effect of long term exposure was observed by McCracken et al. (2010) in the cohort of Veteran Affairs Normative Aging Study (Dioni et al. 2011, McCracken et al. 2010).

Similar research was carried out by Shan et al. (2014). They conducted a cross-sectional study, recruited 25 women in Sichuan, China, and assessed the effect of combustion in house kitchens on PM2.5, inflammation and telomere length. They showed that highly exposed women had 43% shorter telomeres than women with low exposure; however, the sample size was very small (Shan et al. 2014).

Hoxha et al. (2009) assessed the effect of exposure to traffic pollution on traffic officers and indoor office workers (as control group). They also measured the level of benzene and toluene in the air samples. They found that telomere length of traffic officers working in highly polluted area (high traffic intensity) was significantly reduced compared to office workers, and was negatively associated with measured levels of benzene and toluene. They found that traffic officers had telomere length comparable to telomere length observed in 10-years older controls (Hoxha et al. 2009).

Probably due to heterogeneity of measured exposure the results are inconsistent and further studies are needed.

Persistent organic pollutants (POPs)

Persistent organic pollutants constitute a heterogeneous group of lipid-soluble organic chemicals, which are resistant to degradation and prone to bioaccumulation. In that heterogeneous group, one can distinguish polychlorinated biphenyls (non-dioxin like), dioxins, some organochlorine pesticides (like aldrin, mirex, toxaphen, hexachlorobenzene, dieldrin, endrin, chlordane, heptachlor and dichlorodiphenyltrichloroethane—known as DDT) and polychlorinated dibenzofurans. They cause adverse effects in humans and are known due to their endocrine disruptor properties. POPs are classified as the IARC group 1—carcinogenic to humans.

There were 3 population based cross-sectional studies with consistent results. Shin et al. (2010) studied POPs in 84 healthy Koreans and showed that most organochlorine pesticides and polychlorinated biphenyls were associated with longer telomere length (Shin et al. 2010). Similar results were described by Mitro et al. (2015) and Scinicariello and Buser (2015) in NHANES cohort. Mitro et al. (2015) showed that POPs with high toxic equivalency factor and high affinity to bind to aryl hydrocarbon receptor (PCB-126 and PCB-169) were associated with longer telomeres, and Scinicariello and Buser (2015) showed that the sum of PCBs was associated with longer telomeres in environmental exposure and additionally, 1,2,3,4,6,7,8-heptachlorodibenzofuran and 1,2,3,6,7,8-hexachlorodibenzo-p-dioxin were associated with longer telomeres (Mitro et al. 2015, Scinicariello and Buser 2015).

However, results of *in vitro* studies by Jacobus et al. (2008) and Xin et al. (2016), as well as a prospective study by Hou et al. (2013) on pesticide-exposed workers are opposite (Table 1 and Table 2) (Jacobus et al. 2008, Hou et al. 2013, Xin et al. 2016).

Pesticides

Pesticides are chemical compounds which exert biocide action mainly against insects, fungi, microbes, molluscs, rodents and others. They act as plant protection products. There is a huge diversity in chemical structure of pesticides, therefore they consist a heterogeneous group. However, in current studies, it appears that humans exposed to pesticides had shorter telomeres (Table 1). Studies were performed in adults occupationally exposed and the data of exposures based upon questionnaires (Andreotti et al. 2015, Hou et al. 2013, Kahl et al. 2016). Further studies are needed particularly in environmentally exposed populations.

Psychosocial stress

Psychosocial stress, traumatic experiences and social deprivation, however difficult for measurement, also may lead to telomere shortening (Table 3). During the last century, lifestyle and employment has changed. Humans experience new phenomena of stress during work like mobbing, "rat race" and other occurrences, negatively affecting work and living.

Parks et al. (2011) conducted a research in Sister Study cohort. They examined how work affected telomere length in 608 women. It occurred that women with moderate or substantial past work history (more than 20 years full time comparing to those working 1–5 years) or those currently working full-time had shorter telomeres (Parks et al. 2011).

Social disadvantage, recognized as chronic stress, may result from low family income, poverty, low maternal education, unstable family structure and harsh parenting. Such experiences may influence on health and telomere length. Nine-years old boys who were brought up in harsh environment had significantly shorter by 19% telomeres than their colleagues who were brought up in nurturing environment (Mitchell et al. 2014).

Telomere reduction due to psychosocial stress may be long-lasting. Tyrka et al. (2010) examined 31 adults without psychiatric diseases. Subjects were interviewed with Childhood Trauma Questionnaire and those who reported their history of childhood maltreatment had significantly shorter telomeres. The maltreatment examples which were reported were as follows: emotional abuse, emotional neglect, physical abuse, physical neglect, sexual abuse (Tyrka et al. 2010). Also, in adults who experienced childhood adversities (defined as parental unemployment, financial difficulties, parental alcohol problem, mental disease in parents, conflicts in family and school, serious or chronic disease, parents' divorce), accelerated telomere attrition was observed (Kananen et al. 2010). Another stress for a child, which occurs in early life, is being placed in institutional care or foster family. Drury et al. (2012) showed that those children who spent more time in institutional care had shorter telomere than those who were placed in foster care and had a surrogate family (Drury et al. 2012).

Savolainen et al. (2014) assessed telomere length in adults, who were separated from their parents during World War II and those who were not. Isolated separation or reported traumatic experiences were not associated with telomere shortening; however, participants who were separated in childhood and reported traumatic experiences had shorter telomeres when they were senior (Savolainen et al. 2014).

Conclusions

To sum up, both occupational and environmental exposure to chemicals and psychosocial stress result in changes in telomere length. Environmental burden of diseases is still quite high in developed societies. There is no irrefutable proof that telomere shortening or telomere elongation is the direct mechanism by which exposure to single particular chemical results in the specified disease, since most of them are multifactorial. However, telomere length changes are observed both in exposed and in chronically ill populations, and lifestyle or nutritional interventions may be beneficiary and promising in healthy aging of humans, particularly in risk associated populations.

Acknowledgements

The author would like to thank Prof. Julia Bar from Wroclaw Medical University, Poland, for critical review of the manuscript.

This study was supported by the National Science Centre, Poland, DEC-2011/03/D/NZ7/05018.

Keywords: Telomere length, environmental exposure, occupational exposure, psychosocial stress, heavy metal, lead, arsenic, cadmium, pesticides, persistent organic pollutants, dioxins, polychlorinated biphenyls, tobacco smoke, air pollution

References

Andreotti, G., J.A. Hoppin, L. Hou, S. Koutros, S.M. Gadalla, S.A. Savage et al. 2015. Pesticide use and relative leukocyte telomere length in the Agricultural Health Study. Plos One. 10(7): e0133382.

Babizhayev, M.A. and Y.E. Yegorov. 2011. Smoking and health: association between telomere length and factors impacting on human disease, quality of life and life span in a large population-based cohort under the effect of smoking duration. Fundam. Clin. Pharmacol. 25(4): 425–442.

Bassig, B.A., L. Zhang, R.M. Cawthon, M.T. Smith, S. Yin, G. Li et al. 2014. Alterations in leukocyte telomere length in workers occupationally exposed to benzene. Environ. Mol. Mutagen. 55: 673–678.

Cassidy, A., I. De Vivo, Y. Liu, J. Han, J. Prescott, D.J. Hunter et al. 2010. Associations between diet, lifestyle factors and telomere length in women. Am. J. Clin. Nutr. 91: 1273–1280.

Cawthon, R.M. 2002. Telomere measurement by quantitative PCR. Nucleic Acids Res. 30(10) e47: 1–6.

CDC. 2009. Centers for Disease Control and Prevention. Fourth National Report on Human Exposure to Environmental Chemicals. Atlanta, GA: U.S. Department of Health and Human Services, Centers for Disease Control and Prevention. http://www.cdc.gov/exposurereport/ (access 17.01.2016).

CDC. 2015. Centers for Disease Control and Prevention. Fourth National Report on Human Exposure to Environmental Chemicals, Updated Tables, (February, 2015). Atlanta, GA: U.S. Department of Health and Human Services, Centers for Disease Control and Prevention.

Dioni, L., M. Hoxha, F. Nordio, M. Bonzini, L. Tarantini, B. Albetti et al. 2011. Effects of short-term exposure to inhalable particulate matter on telomere length, telomerase expression, and telomerase methylation in steel workers. Environ. Health Perspect. 119: 622–627.

Drury, S.S., K. Theall, M.M. Gleason, A.T. Smyke, I. De Vivo, J.Y. Wong et al. 2012. Telomere length and early severe social deprivation: linking early adversity and cellular aging. Mol. Psychiatry. 17(7): 719–727.

EEA/JCR. 2013. European Environment Agency and the European Commission's Joint Research Centre. Report EUR 25933 EN Environment and human health joint EEA-JRC report. No. 5. ISSN 1725-9177.

Eshkoor, S.A., F. Jahanshiri, P. Ismail, S.A. Rahman, S. Moin and M.Y. Adon. 2012. Association between telomere shortening and ageing during occupational exposure. J. Med. Biochem. 31: 211–216.

Ferrario, D., A. Collotta, M. Carfi, G. Bowe, M. Vahter, T. Hartung et al. 2009. Arsenic induces telomerase expression and maintains telomere length in human cord blood cells. Toxicology. 260: 132–141.

Fujishiro, K., A.V. Diez-Roux, P. Landsbergis, N.S. Jenny and T. Seeman. 2013. Current emploment status, occupational category, occupational hazard exposure, and job stress in relation to telomere length: The Multiethnic Study of Atherosclerosis (MESA). Occup. Environ. Med. 70(8): 552–560.

Gao, J., S. Roy, L. Tong, M. Argos, F. Jasmine, R. Rahaman et al. 2015. Arsenic exposure, telomere length and expression of telomere-related genes among Bangladeshi individuals. Environ. Res. 136: 462–469.

Garcia-Calzon, S., A. Moleres, M.A. Martinez-Gonzalez, J.A. Martinez, G. Zalba, A. Marti et al. 2015. Dietary total antioxidant capacity is associated with leukocyte telomere length in a children and adolescent population. Clin. Nutr. 34(4): 694–9.

Hecht, S.S. 2002. Tobacco smoke carcinogens and breast cancer. Environ. Mol. Mutagen. 39: 119–126.

Hecht, S.S. and J.M. Samet. 2007. Cigarette smoking. pp. 1519–1551. *In*: W.N. Rom and S.B. Markowitz [eds.]. Environmental and Occupational Medicine Fourth Edition. Wolters Kluwer/Lippincott Williams and Wilkins, Philadephia, USA.

Hou, L., S. Wang, C. Dou, X. Zhang, Y. Yu, Y. Zheng et al. 2012. Air pollution exposure and telomere length in highly exposed subjects in Beijing, China: a repeated-measure study. Environ. Int. 48: 71–77.

Hou, L., G. Andreotti, A.A. Baccarelli, S. Savage, J.A. Hoppin, D.P. Sandler et al. 2013. Lifetime pesticide use and telomere shortening among male pesticide applicators in the agricultural health study. Environ. Health Perspect. 121: 919–924.

Hou, L., B.T. Joyce, T. Gao, L. Liu, Y. Zheng, F.J. Penedo et al. 2015. Blood telomere length attrition and cancer development in the Normative Aging Study Cohort. EbioMedicine. 2: 591–596.

Hoxha, M., L. Dioni, M. Bonzini, A.C. Pesatori, S. Fustinoni, D. Cavallo et al. 2009. Association between leukocyte telomere shortening and exposure to traffic pollution: a cross-sectional study on traffic officers and indoor office workers. Environ. Health. 8: 41.

Jacobus, J.A., S. Flor, A. Klingelhutz, L.W. Robertson and G. Ludewig. 2008. 2-(4'-chlorophenyl)-1,4-benzoquinone increases the frequency of micronuclei and shortens telomeres. Environ. Toxicol. Pharmacol. 25(2): 267–272.

Kahl, V.F.S., D. Simon, M. Salvador, C. dos Santos Branco, J.F. Diaz, F.R. da Silva et al. 2016. Telomere measurement in individuals occupationally exposed to pesticide mixtures in tobacco fields. Environ. Mol. Mutagen. 57: 74–84.

Kananen, L., I. Surakka, S. Pirkola, J. Suvisaari, J. Lonnqvist, L. Peltonen et al. 2010. Childhood adversities are associated with shorter telomere length at adult age both in individuals with an anxiety disorder and controls. Plos One. 5(5): e10826.

Langie, S.A.S., G. Koppen, D. Desaulniers, F. Al-Mulla, R. Al-Temaimi, A. Amedei et al. 2015. Causes of genome instability: the effect of low dose chemical exposures in modern society. Carcinogenesis. 36, suppl. 1: S61–S88.

Li, H., B.A. Jonsson, C.H. Lindh, M. Albin and K. Broberg. 2011. N-nitrosoamines are associated with shorter telomere length. Scand. J. Work Environ. Health. 37(4): 316–324.

Li, H., K. Engstrom, M. Vahter and K. Broberg. 2012. Arsenic exposure through drinking water is associated with longer telomeres in peripheral blood. Chem. Res. Toxicol. 25: 2333–2339.

Li, H., M. Hedmer, T. Wojdacz, M.B. Hossain, C.H. Lindh, H. Tinnerberg et al. 2015. Oxidative stress, telomere shortening, and DNA methylation in relation to low-to-moderate occupational exposure to welding fumes. Environ. Mol. Mutagen. 56: 684–693.

Li, H., G. Akerman, C. Liden, A. Alhamdow, T.K. Wojdacz, K. Broberg et al. 2016. Alterations of telomere length and DNA methylation in hairdressers: a cross-sectional study. Environ. Mol. Mutagen. 57: 159–167.

Lin, S., X. Huo, Q. Zhang, X. Fan, L. Du, X. Xu et al. 2013. Short placental telomere was associated with cadmium pollution in an electronic waste recycling town in China. PLoS ONE. 8(4): e60815.

Liu, Q., H. Wang, D. Hu, C. Ding, H. Xu and D. Tao. 2004. Effects of trace elements on the telomere lengths of hepatocytes L-02 and hepatoma cells SMMC-7721. Biol. Trace Elem. Res. 100: 215–227.

McCracken, J., A. Baccarelli, M. Hoxha, L. Dioni, S. Melly, B. Coull et al. 2010. Annual ambient black carbon associated with shorter telomeres in elderly men: Veterans affairs normative aging study. Environ. Health Perspect. 118: 1564–1570.

McGrath, M., J.Y.Y. Wong, D. Michaud, D.J. Hunter and I. De Vivo. 2007. Telomere length, cigarette smoking, and bladder cancer risk in men and women. Cancer Epidemiol. Biomarkers Prev. 16(4): 815–819.

Milne, E., N. O'Callaghan, P. Ramankutty, N.H. de Klerk, K.R. Greenop, B.K. Armstrong et al. 2015. Plasma micronutrient levels and telomere length in children. Nutrition. 31: 331–336.

Mitchell, C., J. Hobcraft, S.S. McLanahan, S.R. Siegel, A. Berg, J. Brooks-Gunn et al. 2014. Social disadvantage, genetic sensitivity, and children's telomere lenght. Proc. Natl. Acad. Sci. USA. 111(16): 5944–5949.

Mitro, S.D., L.S. Birnbaum, B.L. Needham and A.R. Zota. 2015. Cross-sectional associations between exposure to persistent organic pollutans and leukocyte telomere lenght among US adults in NHANES, 2001–2002. Environ. Health Perspect. (in press), http://dx.doi.org/10.1289/ehp.1510187.

O'Callaghan, N.J. and M. Fenech. 2011. A quantitative PCR method for measuring absolute telomere length. Biol. Proced. Online. 13: 3.

Pappas, R.S. 2011. Toxic elements in tobacco and in cigarette smoke: inflammation and sensitization. Metallomics. 3(11): 1181–1198.

Parks, C.G., L.A. DeRoo, D.B. Miller, E.C. McCanlies, R.M Cawthon and D.P. Sandler. 2011. Employment and work schedule are related to telomere length in women. Occup. Environ. Med. 68(8): 582–589.

Pawlas, N., A. Płachetka, A. Kozłowska, K. Broberg and S. Kasperczyk. 2015a. Telomere length in children environmentally exposed to low-to-moderate levels of lead. Toxicol. Appl. Pharmacol. 287: 111–118.

Pawlas, N., A. Płachetka, A. Kozłowska, A. Mikołajczyk, A. Kasperczyk, M. Dobrakowski et al. 2015b. Telomere length, telomerase expression and oxidative stress in lead smelters. Toxicol. Ind. Health. (in press), doi:10.1177/0748233715601758.

Pottier, G., M. Viau, M. Ricoul, G. Shim, M. Bellamy, C. Cuceu et al. 2013. Lead exposure induces telomere instability in human cells. Plos One. 8(6): e67501.

Pruss-Ustun, A. and C. Corvalan. 2006. Preventing disease through healthy environments. Towards an estimate of the environmental burden of disease. WHO, Geneva. http://www.who.int/quantifying_ehimpacts/publications/preventingdisease/en/ (access 17.01.2016).

Revesz, D., Y. Milaneschi, J.E. Verhoeven and B.W. Penninx. 2014. Telomere length as a marker of cellular aging is associated with prevalence and progression of metabolic syndrome. J. Clin. Endocrinol. Metab. 99(12): 4607–4615.

Russo, A., F. Modica, S. Guarrera, G. Fiorito, B. Pardini, C. Viberti et al. 2014. Shorter leukocyte telomere length is independently associated with poor survival in patients with bladder cancer. Cancer Epidemiol. Biomarkers Prev. 23(11): 2439–46.

Rustemeier, K., R. Stabbert, H.J. Haussmann, E. Roemer and E.L. Carmines. 2002. Evaluation of the potential effects of ingredients added to cigarettes. Part 2: Chemical composition of mainstream smoke. Food Chem. Toxicol. 40: 93–104.

Salihu, H.M., A. Pradhan, L. King, A. Paothong, C. Nwoga, P.J. Marty et al. 2014. Impact of intrauterine tobacco exposure on fetal telomere length. Am. J. Obstet. Gynecol. 211: 1.e1–1.e8.

Sarkar, P., K. Shiizaki, J. Yonemoto and H. Sone. 2006. Activation of telomerase in BeWo cells by estrogen and 2,3,7,8-tetrachlorodibenzo-p-dioxin in co-operation with c-Myc. Int. J. Oncol. 28: 43–51.

Savolainen, K., J.G. Eriksson, L. Kananen, E. Kajantie, A.K. Pesonen, K. Heinonen et al. 2014. Associations between early life stress, self-reported traumatic experiences across the lifespan and leukocyte telomere length in elderly adults. Biol. Psychol. 97: 35–42.

Scinicariello, F. and M.C. Buser. 2015. Polychlorinated biphenyls and leukocyte telomere length: an analysis of NHANES 1999–2002. EBioMedicine. 2: 1974–1979.

Shan, M., X. Yang, M. Ezzati, N. Chaturvedi, E. Coady, A. Hughes et al. 2014. A feasibility study of the association of exposure to biomass smoke with vascular function, inflammation, and cellular aging. Environ. Res. 135: 165–172.

Shin, J.Y., Y.Y. Choi, H.S. Jeon, J.H. Hwang, S.A. Kim, J.H. Kang et al. 2010. Low-dose persistent organic pollutants increased telomere length in peripheral leukocytes of healthy Koreans. Mutagenesis. 25: 511–516.

Starkweather, A.R., A.A. Alhaeeri, A. Montpetit, J. Brumelle, K. Filler, M. Montpetit et al. 2014. An integrative review of factors associated with telomere length and implications for behavioural research. Nurs. Res. 63(1): 36–50.

Stuart, B.D., J.S. Lee, J. Kozlitina, I. Noth, M.S. Devine, C.S. Glazer et al. 2014. Effect of telomere length on survival in patients with idiopathic pulmonary fibrosis: an observational cohort study with independent validation. Lancet Respir. Med. 2(7): 557–565.

Sun, Q., L. Shi, J. Prescott, S.E. Chiuve, F.B. Hu, I. De Vivo et al. 2012. Healthy lifestyle and leukocyte telomere length in US women. PLoS ONE. 7(5): e38374.

Theall, K.P., S. McKasson, E. Mabile, L.F. Dunaway and S.S. Drury. 2013. Early hits and long-term consequences: tracking the lasting impact of prenatal smoke exposure on telomere length in children. Am. J. Public Health. 103, suppl. 1: S133–S135.

Tyrka, A.R., L.H. Price, H.T. Kao, B. Porton, S.A. Marsella and L.L. Carpenter. 2010. Childhood maltreatment and telomere shortening: preliminary support for an effect of early stress on cellular aging. Biol. Psychiatry. 67(6): 531–534.

Valdes, A.M., T. Andrew, J.P. Gardner, M. Kimura, E. Oelsner, L.F. Cherkas et al. 2005. Obesity, cigarette smoking, and telomere length in women. Lancet. 366: 662–664.

Verde, Z., L. Reinoso-Barbero, L. Chicharro, N. Garatachea, P. Resano, I. Sanchez-Hernandez et al. 2015. Effects of cigarette smoking and nicotine metabolite ratio on leukocyte telomere length. Environ. Res. 140: 488–494.

Xin, X., P.K. Senthilkumar, J.L. Schnoor and G. Ludewig. 2016. Effects of PCB126 and PCB153 on telomerase activity and telomere length in undifferentiated and differentiated HL-60 cells. Environ. Sci. Pollut. Res. 23: 2173–2185.

Wang, Y., X. Duan, Y. Zhang, S. Wang, W. Yao, S. Wang et al. 2015. DNA methylation and telomere damage in occupational people exposed to coal tar pitch. Zhonghua Lao Dong Wei Sheng Zhi Ye Bing Za Zhi. 33(7): 507–511.

Wentzensen, I.M., L. Mirabello, R.M. Pfeiffer and S.A. Savage. 2011. The association of telomere length and cancer: a meta-analysis. Cancer Epidemiol. Biomarkers Prev. 20(6): 1238–1250.

Willeit, P., J. Willeit, A. Brandstatter, S. Ehrlenbach, A. Mayr, A. Gasperi et al. 2010a. Cellular aging reflected by leukocyte telomere length predicts advanced atherosclerosis and cardiovascular disease risk. Atheroscler. Thromb. Vasc. Biol. 30: 1649–1656.

Willeit, P., J. Willeit, A. Mayr, S. Weger, F. Oberhollenzer, A. Brandstatter et al. 2010b. Telomere length and risk of incident cancer and cancer mortality. JAMA. 304(1): 69–75.

Willeit, P., J. Raschenberger, E.E. Heydon, S. Tsimikas, M. Haun, A. Mayr et al. 2014. Leukocyte telomere length and risk of type 2 diabetes mellitus: new prospective cohort study and literature-based meta-analysis. PLoS ONE. 9(11): e112483.

Wong, J.Y., I. De Vivo, X. Lin and D.C. Christiani. 2014a. Cumulative PM2,5 exposure and telomere length in workers exposed to welding fumes. J. Toxicol. Environ. Health A. 77(8): 441–455.

Wong, J.Y., I. De Vivo, X. Lin, S.C. Fang and D.C. Christiani. 2014b. The relationship between inflammatory biomarkers and telomere length in an occupational prospective cohort study. PLoS ONE. 9(1): e87348.

Wong, J.Y., I. De Vivo, X. Lin, R. Grashow, J. Cavallari and D.D. Christiani. 2014c. The association between global DNA methylation and telomere length in a longitudinal study of boilemakers. Genet. Epidemiol. 38: 254–264.

Wu, Y., Y. Liu, N. Ni, B. Bao, C. Zhang and L. Lu. 2012. High lead exposure is associated with telomere length shortening in Chinese battery manufacturing plant workers. Occup. Environ. Med. 69: 557–563.

Zhang, X., S. Lin, W.E. Funk and L. Hou. 2013. Environmental and occupational exposure to chemicals and telomere length in human studies. Occup. Environ. Med. 70: 743–749.

Zota, A.R., B.L. Needham, E.H. Blackburn, J. Lin, S. Kyun Park, D.H. Rehkopf et al. 2015. Associations of cadmium and lead exposure with leukocyte telomere length: findings from National Health and Nutrition Examination Survey, 1999–2002. Am. J. Epidemiol. 181(2): 127–136.

Chapter 7

Telomeres and Physical Activity

Jean Woo,[1,*] *Ruby Yu*[1] and *Nelson Tang*[2]

INTRODUCTION

Telomeres (TL) are protective structures at the end of chromosomes that shortens with each cell replication, until a limit is reached when further cell replication is not possible and the cell dies. TL maintenance depends on the activity of the enzyme telomerase, which is also responsible for increased proliferative potential in cancer cells. There has been much research in the past two decades into factors that affect the rate of TL shortening, and recently into lengthening. TL is related to lifespan and differential life time exposure to oestrogens may explain the longer life expectancy for women. TL is also determined by genetic and environmental factors (Aviv 2006), and plays an important role in the ageing process (Woo et al. 2010). In recent years TL has been included as a panel of markers that are used to quantify biological ageing in young adults (Belsky et al. 2015). While association between TL and various chronic diseases have been documented, the area of increasing interest is how lifestyle patterns (particularly early in life) may shorten or indeed lengthen TL (Shammas 2011, Song et al. 2010). In particular, stress and other psychosocial factors are likely to affect TL (Epel et al. 2004, Starkweather et al. 2014).

Methodological Issues

Peripheral blood leukocyte is the most readily accessible cell type for measurement of TL. And it is generally believed that TL in leukocytes provide good representation of TL of other cells in the body. The classical method of measurement of telomere length is based on Southern blot after restriction digestion of the telomere and thus it

[1] Department of Medicine & Therapeutics, Faculty of Medicine, The Chinese University, Prince of Wales Hospital, Shatin, N.T., Hong Kong.
[2] Department of Chemical Pathology, Faculty of Medicine, The Chinese University of Hong Kong, Prince of Wales Hospital, Shatin, N.T., Hong Kong.
* Corresponding author: jeanwoowong@cuhk.edu.hk

is also called telomere restriction fragment assay (TRF) (Kimura et al. 2010). In TRF assay, TL is measured in absolute length unit of kilo baseparis (kbp). This method was first performed with in-house radioactive probes and now it could be performed using non-radioactive probes by immunochemical signal detection (e.g., Telomere Length Assay from Sigma-Aldrich). TRF is considered as the gold standard assay for TL, therefore subsequently developed methods were calibrated against TRF assay.

While TRF appears to be a validated assay, many fine details need to be observed in order to produce a valid measurement (reviewed recently by Kimura et al. 2010). For example, not only the amount of DNA loaded into gel and the type of gel electrophoresis will affect the results, special attention is also needed in application of suitable image capturing device and computer image analysis. TRF assays have been used by classical research which provided the foundation on understanding of telomere biology (Samani et al. 2001, Valdes et al. 2005, Andrew et al. 2006).

Early studies used the labor intensive TRF method to measure TL (Samani et al. 2001, Valdes et al. 2005). It was immediately realized that TRF method is not suitable for processing hundreds or thousands of samples.

With the advance of real time quantitative PCR (qPCR) during the turn of the century, it quickly became the method of choice for quantitative assay in molecular genetics. Before long, its application extended into measurement of TL. The method proposed by Cawthon (Cawthon 2002) and its subsequent variations were widely adopted in epidemiological studies. It is also known as T/S assay. In brief, two independent qPCR were performed, one by primers specific to telomere (T) and the other specific for a reference gene with single copy in the genome (S), e.g., acidic ribosomal phosphoprotein gene or Beta-globin gene. QPCR provides Ct values, that is the first PCR cycle that a signal is strong enough for detection (stronger than the threshold of detection of the machine). Ct values of the patient samples are standardized against the same reference sample which are carried out independently for the T and S primers. The ratio of the relative quantitation of T to S reactions is also called T/S ratio. This ratio is dependent on the reference sample. If a reference samples with long TL is used, T/S ratio may be found less than 1, indicating that TL of the patient samples were generally shorter than the reference sample. On the other hand, if a reference sample with short TL is used, T/S ratio in the studied group would be higher than 1. T/S ratio is considered proportional to TL measured by TRF method.

After the first report, qPCR method quickly dominated epidemiology studies of TL (Cawthon et al. 2003, Brouilette et al. 2007, Woo et al. 2009, Zhao et al. 2014). Earlier qPCR assays reported good assay properties, including a very low analytical error of less than 6% CV (coefficient of variation). For comparison, analytical error of common immunoassays are in the range of up to 10–15%. Therefore, qPCR was once considered a perfect assay for TL. In contrary, such a low CV was not always the case in our own hands. Although we followed the Minimum Information for Publication of Quantitative Real-Time PCR Experiments (MIQE) (Bustin et al. 2009) and other guidelines for performance of qPCR, the CV of T/S ratio in qPCR may be above 10%. Recently, other researchers also recognized that a CV of less than

10% was not achieved in many epidemiology studies (Hovatta et al. 2012). Recently, it was also found that correlation between T/S ratio and TRF was not linear over the biological range of TL (Aviv et al. 2011, Martin-Ruiz et al. 2014).

Some of these pitfalls had been attribute to samples and technical problems, like quality of DNA template, efficiency of PCR, PCR machine and the analytical software (Aubert et al. 2012, Cunningham et al. 2013, Nussey et al. 2014). Given these limitations and issues about precision of the assay, many investigators rightly used tertile groups or quartile groups in data analysis of TL instead of the raw T/S ratio (Brouilette et al. 2007, Ma et al. 2013). In meta-analysis of data across different study, special normalization approaches will be required (Gardner et al. 2014).

Another category of measurement methods includes quantitative fluorescence *in situ* hybridization (Q-FISH) and FISH analysis by flow cytometry (Flow-FISH) (Baerlocher et al. 2006). Both methods rely on binding of specific fluorescence probe against telomere repeat sequence and measurement of the fluorescence intensity in cells as a representation of the telomere length. The longer the telomere allows more probes to bind and thus gives rise to stronger fluorescence. The principle of the assay is sound and assay precision is good. There is also a special advantage that telomere length can be measured in individual cell (c.f. TRF and qPCR methods only provide an average of TL across the whole population of cells in the sample). With the ability to differentiate various leukocyte subpopulations, Flow-FISH is the only test that can provide telomere length measurement of individual leukocyte subpopulations. However, freshly drawn blood samples have to be analyzed on the same day for Q-FISH and Flow-FISH. This limitation makes these methods difficult to use in epidemiology studies in which most of the samples are stored and processed later in batch. Further information of these assays could also be found in recent reviews (Aubert et al. 2012, Nussey et al. 2014).

Factors Affecting Telomere Length (Fig. 1)

Given the contribution of lifestyle factors to chronic diseases, and the observation that TL shortening is associated with the presence of these diseases (Woo et al. 2010), lifestyle factors may directly affect TL shortening. Lifestyle factors include diet, smoking, physical activity, and alcohol intake. Reports of association with these factors are conflicting, with some studies noting associations with dietary factors (Chan et al. 2010, Cassidy et al. 2010), with smoking (Valdes et al. 2005), with low physical activity (Cherkas et al. 2008), with alcohol intake (Weischer et al. 2014), or with a combination of these factors (Sun et al. 2012), while others have failed to note any associations (Woo et al. 2008). Various factors may account for these conflicting reports, such as methods used for TL assay, types of cells used, methods used to quantify lifestyle factors, and the extent of adjustment for confounding factors in analyses. Lifestyle factors have also been shown to modify the association between chronic diseases and TL. For example healthy lifestyle behavior attenuate the association between TL and coronary atherosclerosis (Diaz et al. 2010).

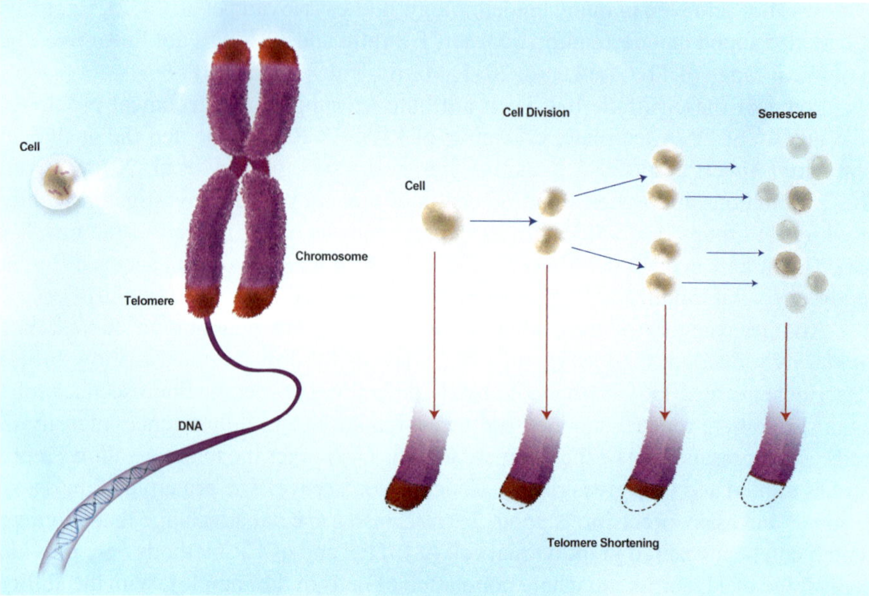

Fig. 1. The process of telomere shortening.

Ageing and Physical Activity

There is extensive evidence that physical activity has a major role in the prevention of chronic diseases of aging. In the recent decade, the concept of frailty has emerged as an important indicator of ageing, incorporating both cognitive as well as physical aspects, moving away from the focus on individual chronic diseases and indeed on multi-morbidity (Harrison et al. 2015, Woo and Leung 2014). Physical activity has been shown to have beneficial effects in prevention of the frail state, and in transition to higher levels of frailty (Cesari et al. 2015, Clegg et al. 2014).

The beneficial effects of physical activity operate via many pathways, such as muscle anabolism in countering sarcopenia (Cruz-Jentoft et al. 2014), improving cardiorespiratory fitness and cognitive function (Ngandu et al. 2015). Therefore it is pertinent to examine the association between telomere length and physical activity.

Evidence for association between telomere length and physical activity (Table 1)

The majority of earlier TL studies was observational and cross sectional in nature, measuring TL in peripheral leukocytes. A systematic review up till 2013 showed conflicting results: ten studies showed a positive relationship with TL; nine showed no association; while three showed an inverted U relationship (Ludlow et al. 2013).

While moderate physical activity may be beneficial for the maintenance of TL, strenuous exercise may be detrimental in accelerating TL shortening (Collins et al. 2003, Ludlow et al. 2008), an effect that is not dependent on telomerase activity. The latter appears more related to the hTERT genotype.

One study reports TL in muscle cells, since the relationship between TL and exercise may be more directly examined. In a group of healthy men and women who were physically active, TL was not impaired compared to younger subjects (Ponsot et al. 2008).

In a cohort study of 2006 older men and women aged 65 years and over living in the community, the relationship between leukocyte TL and physical activity was examined with detailed adjustment for various potential confounders including age, gender, body mass index, socioeconomic ranking, smoking, and chronic diseases. No association between TL and physical activity was observed (Woo et al. 2008). It is possible that in older ages, physical activity may have relatively minor contribution to TL; or that other factors such as psychological stress (which had not been included as a co-variate) may be more important. In the Multi-Ethnic Study of Atherosclerosis, leukocyte TL was measured in 981 white, black, and Hispanic men and women aged 45–84 years. Ethnicity, male gender, lower education level, smoking, higher intake of processes meat, were associated with shorter TL; there was only marginal significant association between shorter TL and greater inactive leisure activity (Diez Roux et al. 2009).

Cross sectional cohort studies reporting stronger associations between TL and physical activity are generally younger. In the Nurses Health Study of 7813 women aged 43–70 years, those who were moderately or highly active had longer telomeres compared with those who were least active, adjusting for potential confounding factors (Du et al. 2012). Similarly postmenopausal women with a mean age of 58 years had higher TL in those engaged in habitual exercise (Kim et al. 2012). Among adolescents, ethnicity, gender, and vigorous physical activity were associated with TL (Zhu et al. 2011). In a cohort of 782 men with a mean age of 47, moderate physical activity was associated with longer TL measured after 29 years of follow up after adjusting for cofounding factors, suggesting that physical activity may have a greater impact earlier in the life course (Savela et al. 2013).

Studies using objective measures of physical activity in terms of energy expenditure equivalents tend to show a more consistent relationship between higher exercise energy expenditure and longer TL. A cross sectional study of 944 outpatients with stable coronary heart disease using METs to measure exercise capacity showed that those with low physical fitness have shorter TL (Krauss et al. 2011). Studies in endurance exercises show that endurance-exercise trained older adults do not have shorter TL compared with younger similarly trained adults although older sedentary adults have shorter TL compared with younger sedentary people, suggesting that TL is positively related to maximal aerobic exercise capacity (LaRocca et al. 2010). Ultra marathon runners with a mean age of 43 years have longer TL compared with age matched controls, that translates into a 16 year difference. The difference was not accounted for by differences in cardiovascular risk factors, nor markers of inflammation or adhesion molecules (Denham et al. 2013). However there was no

Table 1. Association between telomere length and physical activity.

Author	Sample size (N), female (%), age (mean ± SD)	Exposure	Telomere sources, TL assay	Results	*P*-values	Key findings
Collins et al. 2003	N = 26, 26.9% FAMS: 42.0 ± 10.2 y Control: 41.2 ± 11.5 y	FAMS	Vastus Lateralis Muscle, Southern blot	Mean TRF length (mean ± SD, kb) by condition: FAMS: 7.8 ± 2.8 Control: 9.6 ± 0.6	0.052 (Unadjusted)	There was no significant difference in the mean TRF lengths of the two groups. But the minimum TRF lengths of the FAMS athletes were significantly shorter than those of the control subjects
				Minimum TRF length (mean ± SD, kb) by condition: FAMS: 4.0 ± 1.8 Control: 5.4 ± 0.6	0.017 (Unadjusted)	
Ludlow et al. 2008	N = 69, 50.7% Overall: 60.3 ± 4.9 y	Self-reported total energy expenditure and exercise energy expenditure (EEE-Kcal/wk)	PBMCs, qPCR	T/S ratio (mean ± SE) by EEE quartile*: EEE Q1: 0.94T 0.03RU EEE Q2: 1.12T 0.03RU EEE Q3: NA EEE Q4: 0.96T 0.03RU	0.001 (Q2 vs. Q1) 0.04 (Q2 vs. Q4) (Unadjusted)	The second EEE quartile exhibited significantly longer TL than both the first and forth EEE quartiles
Ponsot et al. 2008	N = 42, 45.2% Young female: 23.0 ± 4.0 y Young male: 26.0 ± 3.0 y Old female: 75 ± 4.0 y Old male: 75.0 ± 4.0 y	Self-reported regular participation in recreational sports	Tibialis Anterior Muscle, Southern blot	Mean TL (mean ± SD, kb) by age group and gender: Young female: 10.80 ± 0.65 Young male: 10.67 ± 1.12 Old female: 10.50 ± 1.31 Old male: 10.50 ± 0.72	NS (Unadjusted)	Mean and minimum TL values were not significantly different between young and old subject
				Minimum TL (mean ± SD, kb) by age group and gender: Young female: 4.98 ± 0.70 Young male: 5.00 ± 0.70 Old female: 4.75 ± 0.64 Old male: 4.72 ± 0.56	NS (Unadjusted)	

Woo et al. 2008	N = 2006, 51.3% Overall: 72.4 ± 5.1 y	Self-reported activities (PASE score)	Leukocytes, qPCR	Mean TL (mean ± SD, kb) by PASE quartile[1]: PASE Q1: 8.95 ± 0.09 PASE Q2: 9.10 ± 0.09 PASE Q3: 9.16 ± 0.09 PASE Q4: 9.15 ± 0.09	0.31 (Adjusted)	There was no significant difference in mean TL across quartiles of PA
Diez-Roux et al. 2009	N = 981, 52.4% White: 62.6 ± 10.3 y Black: 60.8 ± 10.1 y Hispanics: 61.3 ± 9.7 y	Self-reported inactive leisure activities (MET-min/wk)	Leukocytes, qPCR	Association between TL (T/S ratio) and PA (leisure MET-mins (in 1000s)): Mean difference per unit of PA –0.009 (SE 0.006)	0.1273 (Adjusted)	Marginal significant association between shorter TL and greater inactive leisure activity
Du et al. 2012	N = 7813, 100% Overall: 59 y	Self-reported total time of activities (MET-hr/wk)	Leukocytes, qPCR	Estimated Least-Squares Mean TL (z-score) and 95% CI by PA level: < 1 hour/week: –0.004 (–0.06, 0.06) 1– < 2 hours/week: 0.01 (–0.06, 0.09) 2– < 4 hours/week: 0.10 (0.02, 0.17) 4– < 7 hours/week: 0.07 (–0.02, 0.16) ≥ 7 hours/week: 0.03 (–0.07, 0.12)	0.05 (Adjusted)[2]	Greater moderate- or vigorous-intensity activity was associated with increased TL
Kim et al. 2012	N = 44, 100% Overall: 58.11 ± 6.8 y	Self-reported exercise frequency and duration	PBMCs, qPCR	T/S ratio (mean ± SD) by exercise group: Habitual physical exercise: 0.93 ± 0.25 Sedentary: 0.35 ± 0.09 Association between TL and habitual physical exercise: β = 0.522	< 0.01 (Unadjusted) < 0.01 (Adjusted)	TL was significantly higher in the habitual exercise group than in the sedentary group. In a stepwise multiple regression analysis, habitual exercise was an independent factor associated with the TL of PBMCs in postmenopausal women

Table 1 contd. ...

...Table 1 contd.

Author	Sample size (N), female (%), age (mean ± SD)	Exposure	Telomere sources, TL assay	Results	*P*-values	Key findings
Zhu et al. 2011	N = 667, 51.4% White female: 16.0 ± 1.2 y White male: 16.2 ± 1.2 y Black female: 16.3 ± 1.3 y Black male: 16.1 ± 1.2 y	Self-reported daily movements	Leukocytes, qPCR	Association between TL and vigorous PA: $R^2 = 0.019$	0.009 (Adjusted)	Vigorous PA was positively associated with TL and accounted for 1.9% of the total variance only in girls
Savela et al. 2013	N = 204, 0% Overall: 76 y	Self-reported work-related and leisure-time activities	Leukocytes, Southern blot	Mean TL (mean ± SE, kb) by PA group: Low PA: 8.10 ± 0.07 Moderate PA: 8.27 ± 0.05 High PA: 8.10 ± 0.05 Proportion of short telomeres (% ± SE) by PA group: Low PA: 12.21 ± 0.39 Moderate PA: 11.35 ± 0.25 High PA: 12.39 ± 0.29	 0.03 (Adjusted) 0.02 (Adjusted)	The moderate PA group had longer mean TL than the low, or high PA groups. The proportion of short telomeres was lowest in the moderate PA group and higher in the high, and the low PA groups
Krauss et al. 2011	N = 944	Assessed MET with treadmill test	Leukocytes, qPCR	T/S ratio (mean ± SE) by physical fitness group: Low physical fitness: 0.85 ± 0.02 Moderate physical fitness: 0.86 ± 0.02 High physical fitness: 0.92 ± 0.03	 0.005 (Adjusted)	Lower exercise capacity was significantly associated with shorter mean TL
LaRocca et al. 2010	N = 57, 40.4% Sedentary young (SY): 23 ± 1.0 y Sedentary old (SO): 65 ± 1.0 y Exercise young (EY): 21 ± 1.0 y Exercise old (EO): 62 ± 2.0 y	Self-reported exercises and VO_2 max assessed with treadmill test	Leukocytes, Southern blot	TL (mean ± SD, kb) by age group and exercise group: SY: 8.407 ± 0.218 SO: 7.059 ± 0.141 EY: 8.579 ± 0.413 EO: 7.992 ± 0.169 Correlation of TL and VO_2 max: r = 0.44	 < 0.01 (SO vs. SY) < 0.01 (SO vs. EO) (Unadjusted) < 0.01	TL of the older endurance-trained adults was ~ 0.9 kb greater than their sedentary peers and was not significantly different from young exercise-trained adults

Denham et al. 2013	N = 123, 0% Runners: 43.6 ± 9.2 y Control: 42.8 ± 9.2 y	Self-reported ultra-marathon records	Leukocytes, qPCR	T/S ratio (mean ± SD) by exercise group: Runners: 3.5 ± 0.68 Control: 3.1 ± 0.41	 2.2×10^{-4} (Adjusted)	The ultra-marathon runners had longer TL (T/S ratio) than control independent of cardiovascular risk factors
Mathur et al. 2013	N = 32, 54.4% Athletes: 54 ± 4.0 y Control: 55 ± 5.0 y	Self-reported running history and VO_2 max assessed with treadmill test	Leukocytes, FISH	Lymphocyte TL (mean ± SD) by exercise group: Athletes: 0.975 ± 0.203 Control: 1.011 ± 0.181 Granulocyte TL (mean ± SD) by exercise group: Athletes: 0.894 ± 0.116 Control: 0.897 ± 0.120 Association between TL and VO_2 max: $r^2 = 0.00$	 0.6 (Unadjusted) 0.9 (Unadjusted) NS (Unadjusted)	Athletes and sedentary controls had similar lymphocyte and granulocyte TL. There is no correlation between TL and exercise capacity

CI, Confidence interval; EEE, Total energy expenditure and exercise energy expenditure; FAMS, Fatigued athlete myopathic syndrome; FISH, Fluorescence *in situ* hybridization; kb, Kilo base-pair; MET (hours/week), Metabolic equivalent hours of activity/week; MET (mins/week), Metabolic equivalent minutes of activity/week; MET (mins/day), Metabolic equivalent minutes of activity/day; NA, Not available; NS, Significant; PA, Physical activity; PASE Score, Physical activity scale for the elderly; PBMCs, Peripheral blood mononuclear cells; qPCR, Qualitative polymerase chain reaction; RU, Relative unit; SD, Standard deviation; SE, Standard error of mean; TL, Telomere length; TRF, Telomeric restriction fragment; T/S ratio, Ratio of telomere repeat copy number to single gene copy number; VO_2 max, Maximal oxygen uptake

[1] Q1 denotes lowest physical activity group, Q4 denotes highest physical activity group.

[2] *P*-value for trend.

association between physical activity measured by maximal oxygen uptake and TL in 17 marathon runners with a mean age of 54 years compared with 15 age matched controls (Mathur et al. 2013).

The underlying mechanism for the association between TL and physical activity is uncertain. Increase in telomerase activity may be a factor, as peripheral blood leukocytes from long term endurance training athletes showed increase telomerase activity, expression of telomere-stabilizing proteins, and down regulation of cell-cycle inhibitors compared with untrained individuals (Werner et al. 2009). Acute exercise has also been observed to lead to upregulation of telomerase reverse transcriptase mRNA and sirtuin-6 mRNA expression (Chilton et al. 2014). Other factors may operate directly on via inflammatory, oxidative, and stress hormonal pathways (Puterman et al. 2010).

There are comparatively fewer studies examining changes in TL with time, either as an observational study or as an end point in randomized controlled trials of various intervention programmes. A study of TL in human skeletal muscle examining the effects of aging and physical activity suggests that telomeres in skeletal muscle are dynamic structures that may be influenced by the environment to be shortened, maintained, or even lengthened (Kadi and Ponsot 2010). A follow-up of 768 men and women over six years showed that telomere shortening is associated with progression of subclinical atherosclerosis using carotid artery intimal thickness as a surrogate marker (Baragetti et al. 2015). On the other hand several studies show that TL can be lengthened by lifestyle modification. Three weeks of intensive lifestyle modification among men with low risk prostatic cancer consisting of dietary regimen, exercises and stress reduction programmes increased telomerase activity in peripheral blood mononuclear cells (Ornish et al. 2008). Subsequently after a three month period of lifestyle intervention, adherence to lifestyle changes was significantly associated with relative TL after adjusting for age and length of the follow-up period (Ornish et al. 2013). Among a group of cervical cancer survivors, counselling intervention which improved patients' quality of life and modulated stress-associated hormones and immune status also resulted in lengthening of telomeres (Peres 2011). In a 12 month study of TL changes among divers exposed to hyperbaric oxygen and aged matched controls where both groups undertook extreme physical activity, TL in granulocytes increased in both groups, but elongation was more marked in granulocytes and naïve T cells in divers (Shlush et al. 2011). Increased telomerase activity was proposed to be an underlying mechanism. A 4.5 year follow up of a Finnish diabetic cohort participating in a lifestyle intervention programme showed that TL increased in about two-thirds of the participants in both intervention and control groups (Hovatta et al. 2012). Weight loss intervention among 74 obese adolescents was accompanied by a significant increase in TL (Garcia-Calzon et al. 2014).

Conflicting results were reported from some studies. In a study of changes in oxidant and anti-oxidant levels with exercise training in obese middle-aged women, TL did not change significantly after acute exercise test inspite of increased oxidative stress (Shin et al. 2008). Among postmenopausal women, 12 months of dietary weight loss and exercise did not change TL. However baseline TL was inversely associated with change in TL (Mason et al. 2013). A large population study of 4576 individuals from the general population in Denmark was followed up for ten years. TL became shorter in 56% and longer in 44%. TL change was inversely associated with baseline TL and age at baseline but not with presence of chronic diseases, lifestyle factors including physical activity, nor with mortality (Weischer et al. 2014).

Conclusion

Human studies so far suggest a protective effect of physical activity on TL shortening. However results are still conflicting, and may be a result of use of different methods to determine TL, as well as differences in study design. There is consensus that the underlying mechanisms include genetic and environmental influence on telomere biology, inflammatory, oxidative and hormonal pathways and their interaction. These remain to be studied in greater detail, in addition to the response of muscle specific telomere.

Keywords: Telomeres, ageing, telomere length, physical activity, lifestyle

References

Andrew, T., A. Aviv, M. Falchi, G.L. Surdulescu, J.P. Gardner, X. Lu et al. 2006. Mapping genetic loci that determine leukocyte telomere length in a large sample of unselected female sibling pairs. American Journal of Human Genetics. 78(3): 480–486. doi:10.1086/500052.

Aubert, G., M. Hills and P.M. Lansdorp. 2012. Telomere length measurement-caveats and a critical assessment of the available technologies and tools. Mutation Research. 730(1-2): 59–67. doi:10.1016/j.mrfmmm.2011.04.003.

Aviv, A. 2006. Telomeres and human somatic fitness. The Journals of Gerontology Series A, Biological Sciences and Medical Sciences. 61(8): 871–873.

Aviv, A., S.C. Hunt, J. Lin, X. Cao, M. Kimura and E. Blackburn. 2011. Impartial comparative analysis of measurement of leukocyte telomere length/DNA content by Southern blots and qPCR. Nucleic Acids Research. 39(20): e134. doi:10.1093/nar/gkr634.

Baerlocher, G.M., I. Vulto, G. de Jong and P.M. Lansdorp. 2006. Flow cytometry and FISH to measure the average length of telomeres (flow FISH). Nat. Protoc. 1(5): 2365–2376.

Baragetti, A., J. Palmen, K. Garlaschelli, L. Grigore, F. Pellegatta, E. Tragni et al. 2015. Telomere shortening over 6 years is associated with increased subclinical carotid vascular damage and worse cardiovascular prognosis in the general population. Journal of Internal Medicine. 277(4): 478–487. doi:10.1111/joim.12282.

Belsky, D.W., A. Caspi, R. Houts, H.J. Cohen, D.L. Corcoran, A. Danese et al. 2015. Quantification of biological aging in young adults. Proceedings of the National Academy of Sciences of the United States of America. 112(30): E4104–4110. doi:10.1073/pnas.1506264112.

Brouilette, S.W., J.S. Moore, A.D. McMahon, J.R. Thompson, I. Ford, J. Shepherd et al. 2007. Telomere length, risk of coronary heart disease, and statin treatment in the West of Scotland Primary Prevention Study: a nested case-control study. Lancet. 369(9556): 107–114.

Bustin, S.A., V. Benes, J.A. Garson, J. Hellemans, J. Huggett, M. Kubista et al. 2009. The MIQE guidelines: minimum information for publication of quantitative real-time PCR experiments. Clinical Chemistry. 55(4): 611–622. doi:10.1373/clinchem.2008.112797.

Cassidy, A., I. De Vivo, Y. Liu, J. Han, J. Prescott, D.J. Hunter et al. 2010. Associations between diet, lifestyle factors, and telomere length in women. The American Journal of Clinical Nutrition. 91(5): 1273–1280. doi:10.3945/ajcn.2009.28947.

Cawthon, R.M. 2002. Telomere measurement by quantitative PCR. Nucleic Acids Research. 30(10): e47.

Cawthon, R.M., K.R. Smith, E. O'Brien, A. Sivatchenko and R.A. Kerber. 2003. Association between telomere length in blood and mortality in people aged 60 years or older. Lancet. 361(9355): 393–395.

Cesari, M., B. Vellas, F.C. Hsu, A.B. Newman, H. Doss, A.C. King et al. 2015. A physical activity intervention to treat the frailty syndrome in older persons-results from the LIFE-P study. The

Journals of Gerontology Series A, Biological Sciences and Medical Sciences. 70(2): 216–222. doi:10.1093/gerona/glu099.

Chan, R., J. Woo, E. Suen, J. Leung and N. Tang. 2010. Chinese tea consumption is associated with longer telomere length in elderly Chinese men. The British Journal of Nutrition. 103(1): 107–113. doi:10.1017/S0007114509991383.

Cherkas, L.F., J.L. Hunkin, B.S. Kato, J.B. Richards, J.P. Gardner, G.L. Surdulescu et al. 2008. The association between physical activity in leisure time and leukocyte telomere length. Archives of Internal Medicine. 168(2): 154–158. doi:10.1001/archinternmed.2007.39.

Chilton, W.L., F.Z. Marques, J. West, G. Kannourakis, S.P. Berzins, B.J. O'Brien et al. 2014. Acute exercise leads to regulation of telomere-associated genes and microRNA expression in immune cells. PloS One. 9(4): e92088. doi:10.1371/journal.pone.0092088.

Clegg, A., S. Barber, J. Young, S. Iliffe and A. Forster. 2014. The home-based older people's exercise (HOPE) trial: a pilot randomised controlled trial of a home-based exercise intervention for older people with frailty. Age and Ageing. 43(5): 687–695. doi:10.1093/ageing/afu033.

Collins, M., V. Renault, L.A. Grobler, A. St. Clair Gibson, M.I. Lambert, E. Wayne Derman et al. 2003. Athletes with exercise-associated fatigue have abnormally short muscle DNA telomeres. Medicine and Science in Sports and Exercise. 35(9): 1524–1528. doi:10.1249/01.MSS.0000084522.14168.49.

Cruz-Jentoft, A.J., F. Landi, S.M. Schneider, C. Zuniga, H. Arai, Y. Boirie et al. 2014. Prevalence of and interventions for sarcopenia in ageing adults: a systematic review. Report of the International Sarcopenia Initiative (EWGSOP and IWGS). Age and Ageing. 43(6): 748–759. doi:10.1093/ageing/afu115.

Cunningham, J.M., R.A. Johnson, K. Litzelman, H.G. Skinner, S. Seo, C.D. Engelman et al. 2013. Telomere length varies by DNA extraction method: implications for epidemiologic research. Cancer Epidemiol. Biomarkers Prev. 22(11): 2047–2054.

Denham, J., C.P. Nelson, B.J. O'Brien, S.A. Nankervis, M. Denniff, J.T. Harvey et al. 2013. Longer leukocyte telomeres are associated with ultra-endurance exercise independent of cardiovascular risk factors. PloS One. 8(7): e69377. doi:10.1371/journal.pone.0069377.

Diaz, V.A., A.G. Mainous 3rd, C.J. Everett, U.J. Schoepf, V. Codd and N.J. Samani. 2010. Effect of healthy lifestyle behaviors on the association between leukocyte telomere length and coronary artery calcium. The American Journal of Cardiology. 106(5): 659–663. doi:10.1016/j.amjcard.2010.04.018.

Diez Roux, A.V., N. Ranjit, N.S. Jenny, S. Shea, M. Cushman, A. Fitzpatrick et al. 2009. Race/ethnicity and telomere length in the Multi-Ethnic Study of Atherosclerosis. Aging Cell. 8(3): 251–257. doi:10.1111/j.1474-9726.2009.00470.x.

Du, M., J. Prescott, P. Kraft, J. Han, E. Giovannucci, S.E. Hankinson et al. 2012. Physical activity, sedentary behavior, and leukocyte telomere length in women. American Journal of Epidemiology. 175(5): 414–422. doi:10.1093/aje/kwr330.

Epel, E.S., E.H. Blackburn, J. Lin, F.S. Dhabhar, N.E. Adler, J.D. Morrow et al. 2004. Accelerated telomere shortening in response to life stress. Proceedings of the National Academy of Sciences of the United States of America. 101(49): 17312–17315. doi:10.1073/pnas.0407162101.

Garcia-Calzon, S., A. Moleres, A. Marcos, C. Campoy, L.A. Moreno, M.C. Azcona-Sanjulian et al. 2014. Telomere length as a biomarker for adiposity changes after a multidisciplinary intervention in overweight/obese adolescents: the EVASYON study. PloS One. 9(2): e89828. doi:10.1371/journal.pone.0089828.

Gardner, M., D. Bann, L. Wiley, R. Cooper, R. Hardy, D. Nitsch et al. 2014. Gender and telomere length: systematic review and meta-analysis. Experimental Gerontology. 51: 15–27.

Harrison, J.K., A. Clegg, S.P. Conroy and J. Young. 2015. Managing frailty as a long-term condition. Age and Ageing. 44(5): 732–735. doi:10.1093/ageing/afv085.

Hovatta, I., V.D. de Mello, L. Kananen, J. Lindstrom, J.G. Eriksson, P. Ilanne-Parikka et al. 2012. Leukocyte telomere length in the finnish diabetes prevention study. PloS One. 7(4): e34948. doi:10.1371/journal.pone.0034948.

Kadi, F. and E. Ponsot. 2010. The biology of satellite cells and telomeres in human skeletal muscle: effects of aging and physical activity. Scandinavian Journal of Medicine & Science in Sports. 20(1): 39–48. doi:10.1111/j.1600-0838.2009.00966.x.

Kim, J.H., J.H. Ko, D.C. Lee, I. Lim and H. Bang. 2012. Habitual physical exercise has beneficial effects on telomere length in postmenopausal women. Menopause. 19(10): 1109–1115. doi:10.1097/gme.0b013e3182503e97.

Kimura, M., R.C. Stone, S.C. Hunt, J. Skurnick, X. Lu, X. Cao et al. 2010. Measurement of telomere length by the Southern blot analysis of terminal restriction fragment lengths. Nat. Protoc. 5(9): 1596–1607.

Krauss, J., R. Farzaneh-Far, E. Puterman, B. Na, J. Lin, E. Epel et al. 2011. Physical fitness and telomere length in patients with coronary heart disease: findings from the Heart and Soul Study. PloS One. 6(11): e26983. doi:10.1371/journal.pone.0026983.

LaRocca, T.J., D.R. Seals and G.L. Pierce. 2010. Leukocyte telomere length is preserved with aging in endurance exercise-trained adults and related to maximal aerobic capacity. Mechanisms of Ageing and Development. 131(2): 165–167. doi:10.1016/j.mad.2009.12.009.

Ludlow, A.T., J.B. Zimmerman, S. Witkowski, J.W. Hearn, B.D. Hatfield and S.M. Roth. 2008. Relationship between physical activity level, telomere length, and telomerase activity. Medicine and Science in Sports and Exercise. 40(10): 1764–1771. doi:10.1249/MSS.0b013e31817c92aa.

Ludlow, A.T., L.W. Ludlow and S.M. Roth. 2013. Do telomeres adapt to physiological stress? Exploring the effect of exercise on telomere length and telomere-related proteins. BioMed. Research International. 2013: 601368. doi:10.1155/2013/601368.

Ma, S.L., E.S. Lau, E.W. Suen, L.C. Lam, P.C. Leung, J. Woo et al. 2013. Telomere length and cognitive function in southern Chinese community-dwelling male elders. Age and Ageing. 42(4): 450–455. doi:10.1093/ageing/aft036.

Martin-Ruiz, C.M., D. Baird, L. Roger, P. Boukamp, D. Krunic, R. Cawthon et al. 2014. Reproducibility of telomere length assessment: an international collaborative study. International Journal of Epidemiology. doi:10.1093/ije/dyu191.

Mason, C., R.A. Risques, L. Xiao, C.R. Duggan, I. Imayama, K.L. Campbell et al. 2013. Independent and combined effects of dietary weight loss and exercise on leukocyte telomere length in postmenopausal women. Obesity (Silver Spring). 21(12): E549–554. doi:10.1002/oby.20509.

Mathur, S., A. Ardestani, B. Parker, J. Cappizzi, D. Polk and P.D. Thompson. 2013. Telomere length and cardiorespiratory fitness in marathon runners. Journal of Investigative Medicine: The Official Publication of the American Federation for Clinical Research. 61(3): 613–615. doi:10.231/JIM.0b013e3182814cc2.

Ngandu, T., J. Lehtisalo, A. Solomon, E. Levalahti, S. Ahtiluoto, R. Antikainen et al. 2015. A 2 year multidomain intervention of diet, exercise, cognitive training, and vascular risk monitoring versus control to prevent cognitive decline in at-risk elderly people (FINGER): a randomised controlled trial. Lancet. 385(9984): 2255–2263. doi:10.1016/S0140-6736(15)60461-5.

Nussey, D.H., D. Baird, E. Barrett, W. Boner, J. Fairlie, N. Gemmell et al. 2014. Measuring telomere length and telomere dynamics in evolutionary biology and ecology. Methods in Ecology and Evolution/British Ecological Society. 5(4): 299–310. doi:10.1111/2041-210X.12161.

Ornish, D., J. Lin, J. Daubenmier, G. Weidner, E. Epel, C. Kemp et al. 2008. Increased telomerase activity and comprehensive lifestyle changes: a pilot study. The Lancet Oncology. 9(11): 1048–1057. doi:10.1016/S1470-2045(08)70234-1.

Ornish, D., J. Lin, J.M. Chan, E. Epel, C. Kemp, G. Weidner et al. 2013. Effect of comprehensive lifestyle changes on telomerase activity and telomere length in men with biopsy-proven low-risk prostate cancer: 5-year follow-up of a descriptive pilot study. The Lancet Oncology. 14(11): 1112–1120. doi:10.1016/S1470-2045(13)70366-8.

Peres, J. 2011. Telomere research offers insight on stress-disease link. Journal of the National Cancer Institute. 103(11): 848–850. doi:10.1093/jnci/djr214.

Ponsot, E., J. Lexell and F. Kadi. 2008. Skeletal muscle telomere length is not impaired in healthy physically active old women and men. Muscle & Nerve. 37(4): 467–472. doi:10.1002/mus.20964.

Puterman, E., J. Lin, E. Blackburn, A. O'Donovan, N. Adler and E. Epel. 2010. The power of exercise: buffering the effect of chronic stress on telomere length. PloS One. 5(5): e10837. doi:10.1371/journal.pone.0010837.

Samani, N.J., R. Boultby, R. Butler, J.R. Thompson and A.H. Goodall. 2001. Telomere shortening in atherosclerosis. Lancet. 358(9280): 472–473. doi:10.1016/S0140-6736(01)05633-1.

Savela, S., O. Saijonmaa, T.E. Strandberg, P. Koistinen, A.Y. Strandberg, R.S. Tilvis et al. 2013. Physical activity in midlife and telomere length measured in old age. Experimental Gerontology. 48(1): 81–84. doi:10.1016/j.exger.2012.02.003.

Shammas, M.A. 2011. Telomeres, lifestyle, cancer, and aging. Current Opinion in Clinical Nutrition and Metabolic Care. 14(1): 28–34. doi:10.1097/MCO.0b013e32834121b1.

Shin, Y.A., J.H. Lee, W. Song and T.W. Jun. 2008. Exercise training improves the antioxidant enzyme activity with no changes of telomere length. Mechanisms of Ageing and Development. 129(5): 254–260. doi:10.1016/j.mad.2008.01.001.

Shlush, L.I., K.L. Skorecki, S. Itzkovitz, S. Yehezkel, Y. Segev, H. Shachar et al. 2011. Telomere elongation followed by telomere length reduction, in leukocytes from divers exposed to intense oxidative stress—implications for tissue and organismal aging. Mechanisms of Ageing and Development. 132(3): 123–130. doi:10.1016/j.mad.2011.01.005.

Song, Z., G. von Figura, Y. Liu, J.M. Kraus, C. Torrice, P. Dillon et al. 2010. Lifestyle impacts on the aging-associated expression of biomarkers of DNA damage and telomere dysfunction in human blood. Aging Cell. 9(4): 607–615. doi:10.1111/j.1474-9726.2010.00583.x.

Starkweather, A.R., A.A. Alhaeeri, A. Montpetit, J. Brumelle, K. Filler, M. Montpetit et al. 2014. An integrative review of factors associated with telomere length and implications for biobehavioral research. Nursing Research. 63(1): 36–50. doi:10.1097/NNR.0000000000000009.

Sun, Q., L. Shi, J. Prescott, S.E. Chiuve, F.B. Hu, I. De Vivo et al. 2012. Healthy lifestyle and leukocyte telomere length in U.S. women. PloS One. 7(5): e38374. doi:10.1371/journal.pone.0038374.

Valdes, A.M., T. Andrew, J.P. Gardner, M. Kimura, E. Oelsner, L.F. Cherkas et al. 2005. Obesity, cigarette smoking, and telomere length in women. Lancet. 366(9486): 662–664. doi:10.1016/S0140-6736(05)66630-5.

Weischer, M., S.E. Bojesen and B.G. Nordestgaard. 2014. Telomere shortening unrelated to smoking, body weight, physical activity, and alcohol intake: 4,576 general population individuals with repeat measurements 10 years apart. PLoS Genetics. 10(3): e1004191. doi:10.1371/journal.pgen.1004191.

Werner, C., T. Furster, T. Widmann, J. Poss, C. Roggia, M. Hanhoun et al. 2009. Physical exercise prevents cellular senescence in circulating leukocytes and in the vessel wall. Circulation. 120(24): 2438–2447. doi:10.1161/CIRCULATIONAHA.109.861005.

Woo, J., N. Tang and J. Leung. 2008. No association between physical activity and telomere length in an elderly Chinese population 65 years and older. Archives of Internal Medicine. 168(19): 2163–2164. doi:10.1001/archinte.168.19.2163.

Woo, J., E.W. Suen, J.C. Leung, N.L. Tang and S. Ebrahim. 2009. Older men with higher self-rated socioeconomic status have shorter telomeres. Age and Ageing. 38(5): 553–558. doi:10.1093/ageing/afp098.

Woo, J., E. Suen and N.L.S. Tang. 2010. Telomeres and the ageing process. Rev. Clin. Gerontol. 20: 1–9.

Woo, J. and J. Leung. 2014. Multi-morbidity, dependency, and frailty singly or in combination have different impact on health outcomes. Age (Dordr). 36(2): 923–931. doi:10.1007/s11357-013-9590-3.

Zhao, J., Y. Zhu, J. Lin, T. Matsuguchi, E. Blackburn, Y. Zhang et al. 2014. Short leukocyte telomere length predicts risk of diabetes in american indians: the strong heart family study. Diabetes. 63(1): 354–362. doi:10.2337/db13-0744.

Zhu, H., X. Wang, B. Gutin, C.L. Davis, D. Keeton, J. Thomas et al. 2011. Leukocyte telomere length in healthy Caucasian and African-American adolescents: relationships with race, sex, adiposity, adipokines, and physical activity. The Journal of Pediatrics. 158(2): 215–220. doi:10.1016/j.jpeds.2010.08.007.

Chapter 8

Caloric Restriction, Sirtuins, and Ageing

Angelo Avogaro and *Saula Vigili de Kreutzenberg**

INTRODUCTION

The natural lifespan of a species, termed essential lifespan, is the time required to accomplish the Darwinian purpose of life in terms of successful reproduction for the continuation of generations. Genes influence longevity, i.e., the property of being long-lived: however, studies on establishing an association between genes and longevity have reported that the genetic heritability of variance in lifespan is less than 35% (Bostock et al. 2009). An association between human longevity and single nucleotide polymorphisms in a variety of genes in various biological pathways has been observed, including heat-shock response, mitochondrial functions, immune response, and cholesterol metabolism. These genes cover a wide range of biochemical pathways, such as energy metabolism, kinases, kinase receptors, transcription factors, DNA helicases, GTP-binding protein–coupled receptors, chaperones, and cell cycle checkpoint pathways. Whatever the genetic component, however, each one of us may quench the inescapable genetic component by modifying life-style, and hence part of the environment: remarkable is the history of Alvise Cornaro, a Venetian nobleman, known for his books of Discorsi, about the secrets to living long and well with sobriety. He was near to death at the age of 35, when Cornaro modified his eating behaviours, and began to adhere to a calorie restriction diet. He died in Padua at age 98. This is an early, historical, example of what we would call today caloric restriction (CR), and emphasizes the key concept that, although death is an inescapable event, in some ways it can be postponed, or, at least, we can adopt some actions, which are associated with a longer life expectancy. A recently published document from a workshop on ageing prevention concludes that targeting aging will

Department of Medicine, University of Padova, Italy.
Email: angelo.avogaro@unipd.it
* Corresponding author

not just postpone chronic diseases but also prevent multiple age-associated metabolic alterations while extending healthy lifespan (Longo et al. 2015). The final remarks from this group highlight the fact that we cannot separate health-span from longevity, since these two are tightly and innately linked. In this chapter the interrelationships between overfeeding, caloric restriction, and epigenetic effects on human bodies, and their relevance on ageing will be described.

Why We Age?

The theories of aging in humans fall into two main categories: programmed and damage theories (Fig. 1). The programmed theory includes: (1) Programmed Longevity, which states that aging is the result of a sequential switching on and off of certain genes, with senescence being defined as the time when age-associated deficits are manifested; (2) Endocrine Theory, which indicates that biological clocks act through hormones to control the pace of aging (Partridge 2001). Recent studies confirm that aging is hormonally regulated and that the evolutionarily conserved insulin/IGF-1 signalling (IIS) pathway plays a key role in the hormonal regulation of aging; (3) Immunological Theory, which states that the immune system declines over time, which leads to an increased vulnerability to infectious disease and thus aging

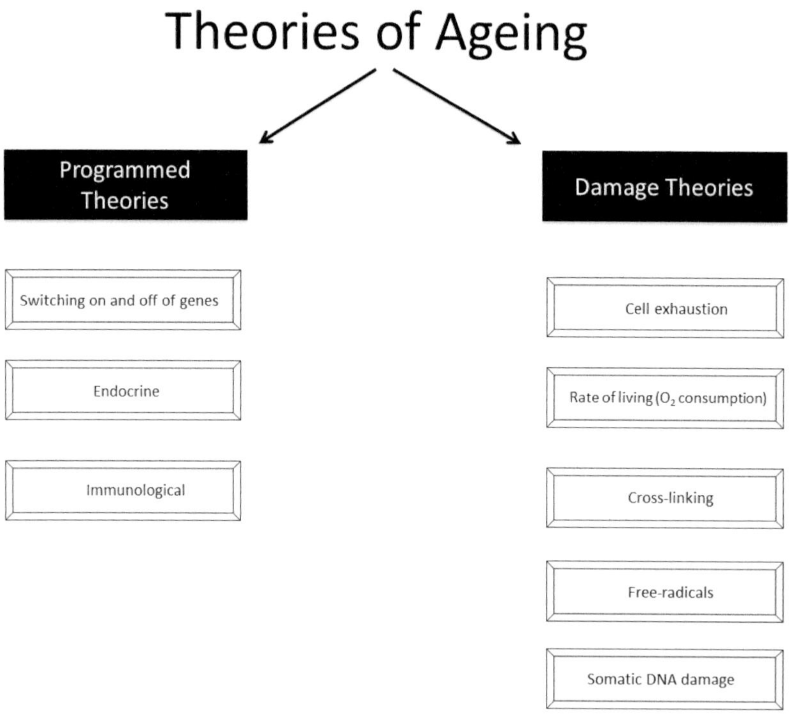

Fig. 1. Theories of ageing.

and death. The damage theory include: (1) Wear and tear theory, which states that cells and tissues have vital parts that exhaust with time resulting in aging; (2) Rate of living theory, which states that the greater the rate of oxygen basal metabolism, the shorter its life span; (3) Cross-linking theory, which indicates that the accumulation of cross-linked proteins damages cells and tissues, slowing down bodily processes resulting in aging; (4) Free radicals theory, which proposes that superoxide and other free radicals cause damage to the macromolecular components of the cell; (5) Somatic DNA damage theory, which supports that as the repair mechanisms cannot correct defects as fast as they are apparently produced, there is a progressive increase in DNA damage accumulation in non-dividing cells of mammals.

Whatever the theory, ageing is considered a progressive decline in the function of multiple cells and tissues: extrinsic threats such as accidents, infections, starvation, and so forth limit the life span of most species. In humans, age-related degeneration gives rise to common pathologies, such, atherosclerosis and heart failure, osteoporosis, pulmonary insufficiency, renal failure, and neurodegeneration. The cellular hallmark of ageing is cellular senescence, i.e., the irreversible arrest of cell proliferation (growth) that occurs when cells experience potentially oncogenic stress: these stimuli include damage to DNA, strong mitogenic signals, damage or disruptions to the epigenome, ectopic expression of certain tumor suppressors (Lopez-Otin et al. 2013). Hayflick and colleague showed that normal human fibroblasts did not proliferate indefinitely in culture (Hayflick 2007): these cells were said to have a finite replicative life span, and, later, to undergo replicative or cellular senescence.

The senescence arrest is considered irreversible because no known physiological stimuli can stimulate senescent cells to re-enter the cell cycle. The senescence state is determined by a mitochondrial dysfunction characterized by a distinct secretory phenotype (SASP) with lower NAD+/NADH ratios, which causes both the growth arrest, and prevents the interleukin (IL)-1-associated SASP through AMPK-mediated p53 activation. SASP exerts different biological activities, the more important being stimulation of angiogenesis, stimulation and inhibition of cell proliferation, creation of a chemoresistant niche during cancer chemotherapy, stimulation of an epithelial-to-mesenchymal transition, chronic inflammation, alterations to stem cell renewal and/or differentiation, and optimization of tissue repair (Campisi 2013). The overall scenario of molecular and cellular senescence is reported in excellent reviews (Erusalimsky and Skene 2009).

Another important accepted theory of senescence is related to telomere biology. Telomere is a region of repetitive nucleotide sequences (TTAGGG is repeated approximately 2,500 times in humans) at each end of a chromosome, which protects the end of the chromosome from deterioration or from fusion with neighbouring chromosomes (Chan and Blackburn 2004). Telomeres progressively shorten in proliferating cells because of two main factors: the "end replication problem" and the oxidative stress. The end replication problem stems from the inability of DNA polymerase to replicate the lagging DNA strand to its terminus, which causes a small and fixed telomere shortening with each replication. In addition, the G triplets of the TTAGGG telomere repeats are susceptible to the oxidative stress. Accordingly, increased oxidative stress causes a greater loss of telomere repeats per cell replication. Heritability estimates of human telomere length range from 30 to 80%.

In humans, associations between stress and telomeres can be seen early in life: newborn human telomere length was shorter in proportion to the stress levels experienced by the mother during her pregnancy (Notterman and Mitchell 2015). A negative effect of exposure to stressors on telomere shortness has also been observed in adults. At a certain critical short length, the cell does not divide further and enters senescence. Cells may circumvent senescence by activating the telomerase complex: one protein in this complex, TERT, synthesizes the TTAGGG repeats of the telomeres through its reverse transcriptase activity. It uses an RNA template produced by TERC, another protein of the telomerase complex. In this way, the telomeric tandem DNA repeats are elongated beyond the critical short length.

Epigenetic modifications in obesity

Hamilton's model predicted that longevity is determined by genes selected for reproductive success, and not by genes that were expressed only late in life (Hamilton 1966). His model showed that the force of natural selection on mortality was highest before the start of the reproduction phase and declined thereafter. Stated in other terms, the force of mortality is highest prior to reproduction: therefore mortality (life span) must be related to genes selected for survival to reproduction age. If modulating ageing by nutritional intervention during old age will not produce the desired results, nutritional approach to obesity should focus among individuals in younger age groups that may predispose them to differential rates of functional decline during advancing age. This is relevant since obesity may lead to epigenetic modification of the genome, i.e., changes in gene expression and cellular phenotypes that are mitotically stable, but that occur without changes in primary DNA sequence are referred to as epigenetic and, at the molecular level, are determined by DNA methylation and histone modifications (Schones et al. 2015). Obesity is a huge health problem of our days. World Health Organization (WHO) estimates that, worldwide, obesity has more than doubled since 1980. In 2014, more than 1.9 billion adults, 18 years and older, were overweight. Of these, over 600 million were obese 42 million children under the age of 5 were overweight or obese in 2013. The fundamental cause of obesity and overweight is an energy imbalance between calories consumed and calories expended. There is a progressive increase in the intake of energy-dense foods that are high in fat, and a decrease in physical inactivity, due to the increasingly sedentary nature of many forms of work, and changing modes of transportation. Nutritional overfeeding during early life generates long-term changes that later predispose to obesity and type 2 diabetes are capable to induce epigenetic modifications of the genome. The transmission of susceptibility to obesity can occur as a consequence of "developmental programming", i.e., the environment encountered at the point of conception and during fetal and neonatal life permanently influences the structure, function, and metabolism of key organs, thus leading to an increased risk of diseases such as type 2 diabetes. Early life exposures are important in promoting adult obesity: critical periods during childhood that appear to influence the subsequent development of obesity, including early infancy, 5–7 years of age (the adiposity rebound period), and puberty (Nicholas et al. 2015). The prenatal stage also represents a window of susceptibility for early life exposures, as offspring of obese humans and animals exhibit increased fat mass very early in life. In differentiated mammalian cells, the principal epigenetic tag found in DNA is that of covalent attachment of a methyl group to the C5 position of cytosine residues in CpG dinucleotide sequences (Zampieri et al. 2015). CpG methylation is, however, an important mechanism to ensure the repression of transcription of repeat elements and transposons, and it plays a crucial role in imprinting and X-chromosome inactivation. One study showed that one of these regions was the BMI-associated region at mass and obesity-associated (FTO) locus FTO, where methylation levels at a CpG site in the first intron of the FTO gene were correlated with genotype at nearby BMI-associated single-nucleotide polymorphisms (Bell et al. 2010). Associations of the FTO risk allele with hypermethylation have now been demonstrated and replicated in several studies (Toperoff et al. 2012, Almen et al. 2014).

Acetylation and deacetylation: role of Sirtuins

Acetylation of the ε-amino group of lysine residues is an additional epigenetic modification, which is known to occur in different organisms for regulating diverse cellular processes. The ϵ-acetylation of lysine residues (Lys-Ac) is an abundant post-translational modification, occurring on thousands of proteins, both nuclear and cytoplasmic, and it plays a regulatory role in all cellular processes (Graff and Tsai 2013). Histones were the first acetylated proteins to be described: for this reason they are still most often referred to as HATs (histone acetyltransferases) and HDACs (histone deacetylases). Histone acetylation at the ε-amino group of lysine residues in H3 and H4 tails is most consistently associated with promotion of transcription. Deacetylation of histones correlates with CpG methylation and the inactive state of chromatin. Acetylation of key lysine residues in H3 and H4 allows a relaxed chromatin structure and active gene transcription: conversely, the recruitment of HDACs to target genes by transcriptional repressors facilitates local histone deacetylation and thus transcriptional repression. The subfamily with the most definitive role in regulating gene expression is the class-1 HDACs: 1, 2, 3 and 8. Class 1 and 2 HDACs have been reported to regulate a variety of metabolic processes, including differentiation of pancreatic islet cells and adipocytes. HDAC5 or HDAC9 gene deletion increased pancreatic β-cell mass and insulin secretion (Lenoir et al. 2011). HDAC3 inhibition was reported to regulate PPARγ acetylation and activity, thereby enhancing insulin signalling and glucose uptake in white adipocytes, and improving insulin sensitivity in diet-induced obese (DIO) mice (Jiang et al. 2014). HDAC9 overexpression represses, whereas HDAC9 gene deletion accelerates, white adipocyte differentiation *in vitro* (Chatterjee et al. 2014). Additionally, HDAC9 deficiency promotes accumulation of beige adipocytes in subcutaneous adipose tissue.

The evolutionary conserved SIR-2 is a NAD+-dependent histone deacetylase; it regulates lifespan in response to caloric restriction in many organisms. Mammalian homologs of SIR2 comprise a family of seven proteins termed sirtuins (SIRT1-SIRT7). These are implicated in metabolic processes and stress resistance (Guarente

2007). SIRT1 represses repetitive DNA, and a functionally diverse set of genes across the mouse genome. In response to DNA damage, SIRT1 dissociates from these loci and re-localizes to DNA breaks to promote repair. Remarkably, increased SIRT1 expression promoted survival in a mouse model of genomic instability, and it suppressed age-dependent transcriptional changes (Oberdoerffer et al. 2008). This occurs through the process of deacetylation, a common reaction that removes an acetyl functional group (CH3-C = O) from a chemical compound. In yeast, SIR2 is a major determinant of longevity because (i) increased SIR2 gene dosage extends lifespan; (ii) loss-of-function mutations shorten it; and (iii) caloric restriction did not extend lifespan when SIR2 is absent. SIRT1 is the closest mammalian homolog of yeast Sir2 and functions as a NAD+-dependent deacetylase. The deacetylation reaction catalyzed by SIRT1 involves the transfer of the acetyl group of an acetylated protein substrate to the ADP-ribose moiety of NAD+, which leads to the generation of a deacetylated protein, 20-acetyl-ADP-ribose and nicotinamide (NAM) (Bordone and Guarente 2005). NAD+ is required for this deacetylation and activates SIRT1 deacetylase activity, while NAM and the reduced form of NAD+ (NADH) inhibit its catalytic activity. SIRT1 modifies the expression of genes involved in cell survival, metabolism, proliferation and differentiation through deacetylation of histones, transcription factors, and transcriptional cofactors. Among the non-histone targets of SIRT1, p53, Foxo, PGC-1a, NCoR/SMRT, LXR, NBS1, and SUV39H1 have been identified: they act as effectors in the SIRT1 signalling controlling metabolism and the response to stress (Fadini et al. 2011). Sirtuins have been reported to favourably regulate key metabolic pathways, including adipogenesis, fatty acid metabolism, amino acid metabolism, and gluconeogenesis. Mice lacking SIRT1 that were fed a high-fat diet developed excessive hepatic lipid accumulation, altered gut microbiota, insulin resistance, and hypertrophic white and brown adipose tissues (Caron et al. 2014). Loss of SIRT3 promoted hyperacetylation of mitochondrial proteins, resulting in metabolic perturbations and increased susceptibility to metabolic disease. SIRT3 was demonstrated to regulate systemic oxidative stress, limit expedited weight gain, and promote metabolic adaptation (McDonnell et al. 2015). Thus, animal studies suggest that SIRTs may protect against the development of metabolic disease. Liver-specific HDAC3 and SIRT6 have been shown to regulate hepatic lipid synthesis, glycolysis, and fatty acid oxidation. Hepatic SIRT1 deficiency also impaired PPARα signalling and decreased fatty acid β-oxidation, resulting in hepatic steatosis, hepatic inflammation, and endoplasmic reticulum stress in high fat–fed mice. Skeletal muscle is also a target of HDACs: class II HDACs have been shown to suppress the formation of type I myofibers, which stimulate insulin-mediated glucose uptake and protect against glucose intolerance (LaBarge et al. 2015). Deletion of SIRT3 in skeletal muscle perturbed mitochondrial function, and promoted oxidative stress and insulin resistance.

Also the vascular endothelium is an important target tissue of the biological functions of SIRT1, where it acts as a modulator of endothelial cells (EC) responses, that integrates metabolic and redox signals into endothelial signalling networks (Avogaro et al. 2013). SIRT1 controls endothelial angiogenic functions and vascular growth: SIRT1 is expressed in ECs of growing blood vessel sprouts. At the cellular level, the inactivation of SIRT1 impairs endothelial migration and causes defects in vessel navigation, whereas its overexpression increases endothelial migration and sprout formation. In ECs SIRT1 binds to Foxo1 to promote its deacetylation, thereby counteracting the antiangiogenic activity of Foxo1 (Oellerich and Potente 2012). In EC other targets of SIRT1 are as the peroxisome proliferator-activated receptor coactivator-1α (PGC-1α), peroxisome proliferator-activated receptor γ (PPARγ) or the endothelial nitric oxide synthase (eNOS). SIRT1 represses PPARγ: activation of PPARγ inhibits EC proliferation and angiogenesis, and SIRT1 has been shown to repress PPARγ target gene expression in adipocytes by binding to its corepressors NCoR/SMRT. SIRT1 can also target eNOS for deacetylation and thereby augment its enzymatic activity (Winnik et al. 2015). eNOS-derived nitric oxide (NO) is essential for endothelium-dependent vasodilatation: therefore it is likely that NO critically contributes to the SIRT1-dependent regulation of vascular growth and homeostasis. SIRT1 not only promotes blood vessel formation, but also enhance vessel dilatation, and possibly may link the state of insulin resistance with the progression of atherosclerotic burden in humans (de Kreutzenberg et al. 2010). In addition SIRT1 affects endothelial angiogenic functions by preventing stress-induced endothelial dysfunction and premature senescence under stressed conditions.

Caloric restriction: mechanistic data

A sedentary lifestyle and overfeeding leads to obesity, which is a powerful risk factor for the metabolic syndrome: this condition leads to an increased risk of type 2 diabetes and cardiovascular disease. In addition, metabolic syndrome is associated with a variety of other systemic complications that affect disparate organs and systems, such as fatty liver disease, respiratory disease, osteoarticular disease, and cancer. As a result, metabolic syndrome patients have increased all-cause mortality and a shortened lifespan compared with the general population. It is also progressively recognized that metabolic syndrome is associated with precocious aging (Nunn et al. 2014). Increased physical activity and reduced caloric intake are the two main cornerstones, along with smoking cessation, of a healthy life-style. It is recognized that dietary caloric restriction (CR) is the only conclusive, and reproducible intervention shown to retard aging and extend healthy life span in mammals (Fontana and Partridge 2015). CR maintains physiological and behavioural functions, and both delays and reduces the severity of many age-related diseases in a variety of animal models. Nutritional factors may induce epigenetic changes via three pathways: (a) a direct influence on gene expression, (b) activation of nuclear receptors by ligands, and (c) modification of membrane receptor signalling cascades. There are many types of dietary interventions that have been shown to affect risk factors for premature ageing: prolonged fasting, intermittent fasting, protein restriction or selective amino acid restriction (Table 1).

CR extends the lifespan of such diverse organisms as yeast, worms, flies, fish, and rodents, and seems likely to have the same effects in nonhuman primates (rhesus monkeys). In a 20-year longitudinal adult-onset calorie restriction study in rhesus macaques maintained at the Wisconsin National Primate Research Center,

Table 1. Effects of nutrients on ageing-related signalling pathways.

Effect	Nutrient
Increased DNA-methylation	folate vitamin B12 selenium epigallocatechin-3-gallate epicatechin ganistein quercetin, fisetin myricetin
Deacetylase inhibition	polyphenols
Histone acetyltransferases	polyphenols copper
Expression of miRNA	anthocyanin, curcumin quercetin genistein retinoic acid fish oil
Stimulation of mTOR pathway	insulin branch-chain aminoacids
Inhibition of mTOR pathway	caffeine rapamycin

the incidence of aging-related deaths was lower in the calorie restriction group (50%) than in the ad libitum group (80%) (Colman et al. 2014). The mechanisms proposed for life-prolongation by CR include an increase in apoptosis, an attenuation of oxidative damage, and a lowering of body temperature. Additional hypothesis emphasize the concept that CR may reduce the endogenous production of damaging agents (such as ROS) as well as the effect of "hormesis", which helps protect the organism from damaging agents and repair any resulting damage when it occurs. The aging process is associated with globally decreased but locally increased DNA methylation: CR is assumed to delay the aging process by reversing aging-related DNA methylation changes thus increasing genomic stability. CR increased the methylation level of proto-oncogene *RAS* in a rat model when compared to *ad libitum* fed animals (Hass et al. 1993). HDAC activity is increased during CR: global deacetylation may therefore be a protective mechanism against nutrition stress and may influence the aging processes. Enhanced activity of HDAC1 on the promoter regions of the $p16^{INK4\alpha}$ and human telomerase reverse transcriptase genes, the former a tumor suppressor in many human cancers, and the latter a key regulator of telomerase activity modified by aging regulation, leads to beneficial expression changes of these two genes and contributes to longevity under CR conditions (Daniel and Tollefsbol 2015). The enzymatic activity of SIRT1 depends on NAD^+/NADH ratio, a key indicator for oxygen consumption, thus suggesting that this protein is responsive to the metabolic state of cells (Imai and Guarente 2014). The role of SIRT1 in mediating

CR and lifespan extension is supported by several animal models, and human subjects (Cohen et al. 2004, Crujeiras et al. 2008). CR induces SIRT1 expression in several tissues of mice or rats: SIRT1 is assumed to mediate CR-induced metabolic alterations and subsequent aging retardation by (a) increasing stress resistance by negative regulation of p53 and FOXO (Luo et al. 2001, Motta et al. 2004) and (b) by initiating a series of endocrine responses, including inhibition of adipogenesis and insulin secretion in pancreatic β cells by regulation of key metabolism-associated genes such as peroxisome proliferator-activated receptor G coactivator 1α (PGC-1α) (Schilling et al. 2006). In contrast to histone acetylation, associated with open chromatin status and subsequent gene activation, differentially methylated forms of histones lead to gene silencing or activation (Sedivy et al. 2008). CR elicited histone methylation modifications such as di- or tri-methylated histone H3 at lysine residue 3 or 4 regulate expression of key aging-related genes, $p16^{INK4\alpha}$ and *hTERT*, thereby contributing to CR-induced lifespan extension of human cells (Li and Tollefsbol 2011).

In addition, CR may positively act through the modulation of insulin/Insulin growth factor(IGF)-1 axis (Avogaro et al. 2010). Bluher and colleagues evaluated life span of the fats specific insulin receptor knockout (FIRKO) mouse: They found that, while growth curves were normal in male and female FIRKO mice from birth to 8 weeks of age, starting at 3 months of age, FIRKO mice maintained 15–25% lower body weights and a 50–70% reduction in fat mass throughout life (Bluher et al. 2003). Quantitative survival curves revealed that the mean life span was increased by 18%. The human counterpart of this observation is the finding that a lower IGF activity has been found in human centenarians (Barbieri et al. 2008). CR induces the secretion of adiponectin by adipose tissue: cross-sectional studies demonstrate a consistent inverse relationship between plasma insulin and adiponectin concentrations (Shinmura et al. 2007). Adiponectin promotes fatty acid oxidation in adipose tissue and reduces lipid accumulation in other peripheral tissues, and regulates mitochondrial energy production via AMPK (Zhu et al. 2007). AMPK has many functions, it up-regulates cellular uptake of glucose, β-oxidation of fatty acids, expression of glucose transporter 4 (GLUT4), and mitochondrial energy production. Increased AMPK activity during CR has also a cardioprotective effect, which is abolished in transgenic adiponectin antisense mice or in mice treated with an AMPK inhibitor (Zhu et al. 2007). Additional cardioprotective effects that are mediated by increased secretion of adiponectin during CR are (a) increase in eNOS activity, (b) inhibition of TNFα release, and (c) inhibition of synthesis of adhesion molecules in endothelial cells.

Life span is regulated also by the mammalian target of rapamycin (mTOR) pathway: inhibition of mTOR with rapamycin expands maximal and median life span in mice (Ruetenik and Barrientos 2015). This effect was observed even when the treatment was initiated late in life, corresponding roughly to an age of 60 years in humans (Harrison et al. 2009). The inhibition of mTORC1, a complex including the classic functions of mTOR and controller of protein synthesis, could prevent tissue degeneration and extend life span by improving stem cell function (Chen et al. 2009). S6K1 and 4E-BP1 were the suggested effectors of the mTORC1 signalling pathway that regulates the aging process: reduced S6K1 activity increases life span in various species including mice (Kapahi et al. 2010). The overexpression of 4E-BP1 extends life span under rich nutrient conditions by enhancing mitochondrial activity in flies (Kapahi et al. 2010). mTORC1 could also influence life span through mechanisms that are not associated with modulation of protein synthesis; for example, stimulation of autophagy, as a consequence of mTORC1 inhibition, could promote longevity by stimulating degradation of aberrant proteins and damaged organelles that are accumulating over time and impairing cellular homeostasis (Hands et al. 2009). An example how alteration of mTORC1 activity can affect life span is seen in the liver of old mice with impaired fasting-induced ketogenesis and increased mTORC1 activity (Sengupta et al. 2010). This impaired ketogenesis limits the supply of available energy substrates to the peripheral tissues thus reducing the organism's chances of survival during food deprivation.

Caloric restriction: Clinical data

Excessive nutrition before birth alters the development of the adipocyte, and it leads to an increase in the ability to form new cells in adipose depots or to store lipids into pre-existing adipocytes. These cells exposed to high glucose and insulin levels before birth exhibit an important increase of lipogenetic capacity, which precedes the development of obesity. Thus, an excess substrate supply during the development precedes an increased capacity for storing lipids in adult life. This predisposes individuals to larger adipose mass, deposition of fat outside adipocytes (so-called ectopic fat), insulin resistance, and eventually to the clinical unwanted causes of this condition, such as a reduced life span (Reusens et al. 2007). The implications of CR for human health and longevity are still under scrutiny: this uncertainty is in part due to the obvious time demands of making observations over human life time, and in part due to changes in adherence that would be evident in a randomized study design, and in part due to dietary qualitative composition. Human subjects voluntarily undergoing calorie restriction are confounded by a number of other essential qualities that characterize the study population (Rizza et al. 2014). Essentially all of the controlled, prospective studies of calorie restriction interventions in humans have been conducted in seriously overweight and obese (BMI $\geq$ 30 kg/m^2) individuals over very short (usually 6 months) durations (Childs et al. 2015). In humans, the relationship between CR and life expectancy is far from being straightforward. There is evidence that the relationship between BMI and survival is "U-shaped", i.e., being underweight as well as overweight is associated with a shortened lifespan. The relationship between BMI and mortality are subject to a number of confounders: an example of this is highlighted by the BODE index (Body-mass index, airflow Obstruction, Dyspnea, and Exercise), a scoring system and capacity index used in patients with chronic obstructive pulmonary disease to predict long-term outcomes: in index, a significant weight loss is associated with a negative outcome. Furthermore, the quality of the diets of those with low BMI may lack important nutrients needed for long life. Notwithstanding these caveats, prospective studies of calorie restriction interventions in humans are available both in non-obese and obese subjects (Table 2).

Table 2. Hormonal and metabolic effects of caloric restriction.

1.	**Reduced fat mass**
2.	**Reduction of IGF1/insulin signalling pathway**
3.	**Increased adiponectin secretion**
4.	**Reduced total cholesterol/HDL ratio**
5.	**Decreased C reactive protein level**
6.	**Reduced blood pressure**
7.	**Reduced chronic inflammation**
8.	**Maintenance of heart rate variability**

Fontana and coworkers evaluated the effect of a 6-year long CR diet on risk factors for atherosclerosis in male and female adults (age range 35–82 years) and compared them to age-matched healthy individuals on typical American diets (control group) (Fontana et al. 2007). The total serum cholesterol level and low-density lipoprotein (LDL) cholesterol levels, the ratio of total cholesterol to high-density lipoprotein cholesterol (HDL), triglycerides, fasting glucose, fasting insulin, C-reactive protein (CRP), platelet-derived growth factor AB, and systolic and diastolic blood pressures were all markedly lower in the CR group. The HDL cholesterol was higher after CR. The conclusion of the study was that long-term CR can reduce the risk factors for atherosclerosis. The effects of a 1-year, 20% CR regime or 20% IEE by exercise, on the oxidative damage of DNA and RNA, were evaluated by white blood cell and urine analyses in normal-to-overweight adults (Hofer et al. 2008). Both interventions significantly reduced oxidative damage to both DNA and RNA in white blood cells compared to baseline. However, urinary levels of DNA and RNA oxidation products did not differ from baseline values following 1-year intervention program. Six months of calorie restriction, with or without exercise, in overweight nonobese men and women, decreases 2 biomarkers of longevity (fasting insulin level and body temperature), and indicates that the metabolic rate is reduced beyond the level expected from reduced metabolic body mass (Heilbronn et al. 2006).

Members of the Calorie Restriction Society (CRS) restrict food intake with the expectation that this would delay the disease processes responsible for secondary aging and to slow the primary aging process (Fontana et al. 2004, Meyer et al. 2006). Compared to age-matched individuals eating typical American diets, CRS members (average age 50 ± 10 yr) had a lower body mass index, a reduced body fat, significantly lower values for total serum cholesterol, LDL cholesterol, total cholesterol/LDL, and higher HDL cholesterol. Also fasting plasma insulin and glucose values were significantly lower than in the age-matched control group. Left ventricular diastolic function in CRS members was similar to that of about 16 years younger individuals. Chronic inflammation was also reduced by CR, and this was reflected in significantly lower levels of plasma C reactive protein and TNFα.

The Comprehensive Assessment of Long-term Effects of Reducing Intake of Energy (CALERIE) trials assess the role of CR in different anthropometric parameters and risk factor for cardiovascular disease (Fontana et al. 2007, Fontana et al. 2016, Villareal et al. 2016, Ravussin et al. 2015, Redman et al. 2014). In particular, 218 nonobese humans aged 21–51 years, were randomized to a 2-year intervention designed to achieve 25% CR or to ad libitum (AL) diet. Outcomes were change from baseline resting metabolic rate adjusted for weight change, core temperature, plasma triiodothyronine, and tumor necrosis factor (TNF)-α (secondary). The CR group achieved a mean 11.7% of CR and maintained a mean 10.4% of weight loss. RMR residual decreased significantly more in CR than AL at 12 months but not 24 months. Core temperature change differed little between groups. T3 and TNF-α decreased more in CR than in AL group. CR group had larger decreases in cardiometabolic risk factors and in daily energy expenditure adjusted for weight change, without adverse effects on quality of life. Finally, aging is associated with a progressive reduction in heart-rate-variability (HRV)—a measure of declining autonomic function—and also a worse health outcome. The effect of a 30% CR on heart autonomic function was assessed by 24-hour monitoring of HRV in adults on self-imposed CR for 3 to 15 years and compared with an age-matched control eating a Western diet (Stein et al. 2012). The CR group had a significantly lower heart rate and significantly higher values for HRV. Also, HRV in the CR individuals was comparable to published norms for healthy individuals 20 years younger. Therefore, CR is able to reset the balance between the sympathetic/parasympathetic modulation of heart frequency in favour of the parasympathetic drive thus increasing the circadian variability of heart rate.

Conclusions

Dietary interventions based on chronic dietary restriction such as periodic fasting mimicking, protein restriction, etc., drugs that inhibit the growth hormone/IGF-I axis, drugs that inhibit the mTOR–S6K pathway, or drugs that activate AMPK or sirtuins have been proposed to delay aging. In the cure of chronic metabolic disease such as obesity and type 2 diabetes, lifestyle plays a crucial role. However, based on Hamilton's assumption stating that longevity is determined by genes selected for reproductive success, and not by genes that were expressed only late in life, the modifications in lifestyle should be implemented early, and not late, in life. Indeed there is evidence that the fetal and early postnatal periods strongly influence the risk of developing obesity in later life: if therefore we believe that this is true, the treatment for a longer life span should be started as early as possible (McDonald and Ruhe 2011) (Fig. 2). Overnutrition in the early life can also increase susceptibility to future obesity and type 2 diabetes mellitus: since diabetes leads to an accelerated atherosclerotic process, we may prevent vascular longevity by tackling these disease beginning from prior generation, and not implementing life style only in late stage of life. Furthermore overnutrition leads to obesity, which begets insulin resistance: this condition is intimately linked to both aging and ED, which is a major driver of CVD and a significant contributor to the accelerated aging process. Decreasing

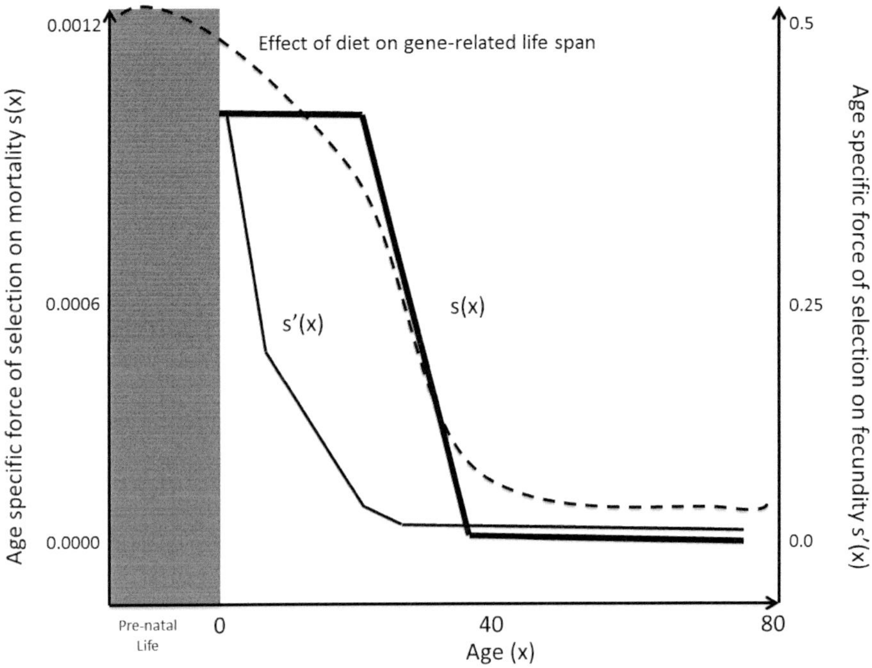

Fig. 2. Effect of diet on gene-related life span according to Hamilton's calculation on the force of natural selection on mortality (s(x)) and fecundity (s'(x)). (Adapted from Hamilton 1966).

insulin resistance will slow endothelial senescence: therefore, combining successful control of modifiable risk factors may thus circumvent the ineluctable power of the genetic background. Targeted intervention on endothelial aging pathways is the next challenge.

Keywords: Ageing, adipose tissue, epigenetic, sirtuin, cardiovascular disease

References

Almen, M.S., E.K. Nilsson, J.A. Jacobsson, I. Kalnina, J. Klovins, R. Fredriksson et al. 2014. Genome-wide analysis reveals DNA methylation markers that vary with both age and obesity. Gene. 548: 61–67.

Avogaro, A., S.V. de Kreutzenberg and G.P. Fadini. 2010. Insulin signaling and life span. Pflugers. Arch. 459: 301–314.

Avogaro, A., S.V. de Kreutzenberg, M. Federici and G.P. Fadini. 2013. The endothelium abridges insulin resistance to premature aging. J. Am. Heart Assoc. 2: e000262.

Barbieri, M., A. Gambardella, G. Paolisso and M. Varricchio. 2008. Metabolic aspects of the extreme longevity. Exp. Gerontol. 43: 74–78.

Bell, C.G., S. Finer, C.M. Lindgren, G.A. Wilson, V.K. Rakyan, A.E. Teschendorff et al. 2010. Integrated genetic and epigenetic analysis identifies haplotype-specific methylation in the FTO type 2 diabetes and obesity susceptibility locus. PLoS One. 5: e14040.

Bluher, M., B.B. Kahn and C.R. Kahn. 2003. Extended longevity in mice lacking the insulin receptor in adipose tissue. Science. 299: 572–574.

Bordone, L. and L. Guarente. 2005. Calorie restriction, SIRT1 and metabolism: understanding longevity. Nat. Rev. Mol. Cell. Biol. 6: 298–305.

Bostock, C.V., R.L. Soiza and L.J. Whalley. 2009. Genetic determinants of ageing processes and diseases in later life. Maturitas. 62: 225–229.

Campisi, J. 2013. Aging, cellular senescence, and cancer. Annu. Rev. Physiol. 75: 685–705.

Caron, A.Z., X. He, W. Mottawea, E.L. Seifert, K. Jardine, D. Dewar-Darch et al. 2014. The SIRT1 deacetylase protects mice against the symptoms of metabolic syndrome. FASEB J. 28: 1306–1316.

Chan, S.R. and E.H. Blackburn. 2004. Telomeres and telomerase. Philos. Trans. R. Soc. Lond. B. Biol. Sci. 359: 109–121.

Chatterjee, T.K., J.E. Basford, K.H. Yiew, D.W. Stepp, D.Y. Hui and N.L. Weintraub. 2014. Role of histone deacetylase 9 in regulating adipogenic differentiation and high fat diet-induced metabolic disease. Adipocyte. 3: 333–338.

Chen, C., Y. Liu, Y. Liu and P. Zheng. 2009. mTOR regulation and therapeutic rejuvenation of aging hematopoietic stem cells. Sci. Signal. 2: ra75.

Childs, B.G., M. Durik, D.J. Baker and J.M. van Deursen. 2015. Cellular senescence in aging and age-related disease: from mechanisms to therapy. Nat. Med. 21: 1424–1435.

Cohen, H.Y., C. Miller, K.J. Bitterman, N.R. Wall, B. Hekking, B. Kessler et al. 2004. Calorie restriction promotes mammalian cell survival by inducing the SIRT1 deacetylase. Science. 305: 390–392.

Colman, R.J., T.M. Beasley, J.W. Kemnitz, S.C. Johnson, R. Weindruch and R.M. Anderson. 2014. Caloric restriction reduces age-related and all-cause mortality in rhesus monkeys. Nat. Commun. 5: 3557.

Crujeiras, A.B., D. Parra, E. Goyenechea and J.A. Martinez. 2008. Sirtuin gene expression in human mononuclear cells is modulated by caloric restriction. Eur. J. Clin. Invest. 38: 672–678.

Daniel, M. and T.O. Tollefsbol. 2015. Epigenetic linkage of aging, cancer and nutrition. J. Exp. Biol. 218: 59–70.

de Kreutzenberg, S.V., G. Ceolotto, I. Papparella, A. Bortoluzzi, A. Semplicini, C. Dalla Man et al. 2010. Downregulation of the longevity-associated protein sirtuin 1 in insulin resistance and metabolic syndrome: potential biochemical mechanisms. Diabetes. 59: 1006–1015.

Erusalimsky, J.D. and C. Skene. 2009. Mechanisms of endothelial senescence. Exp. Physiol. 94: 299–304.

Graff, J. and L.H. Tsai. 2013. Histone acetylation: molecular mnemonics on the chromatin. Nat. Rev. Neurosci. 14: 97–111.

Fadini, G.P., G. Ceolotto, E. Pagnin, S. de Kreutzenberg and A. Avogaro. 2011. At the crossroads of longevity and metabolism: the metabolic syndrome and lifespan determinant pathways. Aging Cell. 10: 10–17.

Fontana, L., T.E. Meyer, S. Klein and J.O. Holloszy. 2004. Long-term calorie restriction is highly effective in reducing the risk for atherosclerosis in humans. Proc. Natl. Acad. Sci. USA. 101: 6659–6663.

Fontana, L., D.T. Villareal, E.P. Weiss, S.B. Racette, K. Steger-May, S. Klein et al. 2007. Calorie restriction or exercise: effects on coronary heart disease risk factors. A randomized, controlled trial. Am. J. Physiol. Endocrinol. Metab. 293: E197–202.

Fontana, L. and L. Partridge. 2015. Promoting health and longevity through diet: from model organisms to humans. Cell. 161: 106–118.

Fontana, L., D.T. Villareal, S.K. Das, S.R. Smith, S.N. Meydani, A.G. Pittas et al. 2016. Effects of 2-year calorie restriction on circulating levels of IGF-1, IGF-binding proteins and cortisol in nonobese men and women: a randomized clinical trial. Aging Cell. 15: 22–27.

Guarente, L. 2007. Sirtuins in aging and disease. Cold Spring Harb. Symp. Quant. Biol. 72: 483–488.

Hamilton, W.D. 1966. The moulding of senescence by natural selection. J. Theor. Biol. 12: 12–45.

Hands, S.L., C.G. Proud and A. Wyttenbach. 2009. mTOR's role in ageing: protein synthesis or autophagy? Aging (Albany NY). 1: 586–597.

Harrison, D.E., R. Strong, Z.D. Sharp, J.F. Nelson, C.M. Astle, K. Flurkey et al. 2009. Rapamycin fed late in life extends lifespan in genetically heterogeneous mice. Nature. 460: 392–395.

Hass, B.S., R.W. Hart, M.H. Lu and B.D. Lyn-Cook. 1993. Effects of caloric restriction in animals on cellular function, oncogene expression, and DNA methylation *in vitro*. Mutat. Res. 295: 281–289.

Hayflick, L. 2007. Biological aging is no longer an unsolved problem. Ann. N. Y. Acad. Sci. 1100: 1–13.

Heilbronn, L.K., L. de Jonge, M.I. Frisard, J.P. DeLany, D.E. Larson-Meyer, J. Rood et al. 2006. Effect of 6-month calorie restriction on biomarkers of longevity, metabolic adaptation, and oxidative stress in overweight individuals: a randomized controlled trial. JAMA. 295: 1539–1548.

Hofer, T., L. Fontana, S.D. Anton, E.P. Weiss, D. Villareal, B. Malayappan et al. 2008. Long-term effects of caloric restriction or exercise on DNA and RNA oxidation levels in white blood cells and urine in humans. Rejuvenation Research. 11: 793–799.

Imai, S. and L. Guarente. 2014. NAD+ and sirtuins in aging and disease. Trends Cell. Biol. 24: 464–471.

Jiang, X., X. Ye, W. Guo, H. Lu and Z. Gao. 2014. Inhibition of HDAC3 promotes ligand-independent PPARgamma activation by protein acetylation. J. Mol. Endocrinol. 53: 191–200.

Kapahi, P., D. Chen, A.N. Rogers, S.D. Katewa, P.W. Li, E.L. Thomas et al. 2010. With TOR, less is more: a key role for the conserved nutrient-sensing TOR pathway in aging. Cell. Metab. 11: 453–465.

LaBarge, S., C. Migdal and S. Schenk. 2015. Is acetylation a metabolic rheostat that regulates skeletal muscle insulin action? Mol. Cells. 38: 297–303.

Lenoir, O., K. Flosseau, F.X. Ma, B. Blondeau, A. Mai, R. Bassel-Duby et al. 2011. Specific control of pancreatic endocrine beta- and delta-cell mass by class IIa histone deacetylases HDAC4, HDAC5, and HDAC9. Diabetes. 60: 2861–2871.

Li, Y. and T.O. Tollefsbol. 2011. p16(INK4a) suppression by glucose restriction contributes to human cellular lifespan extension through SIRT1-mediated epigenetic and genetic mechanisms. PLoS One. 6: e17421.

Longo, V.D., A. Antebi, A. Bartke, N. Barzilai, H.M. Brown-Borg, C. Caruso et al. 2015. Interventions to Slow Aging in Humans: Are We Ready? Aging Cell. 14: 497–510.

Lopez-Otin, C., M.A. Blasco, L. Partridge, M. Serrano and G. Kroemer. 2013. The hallmarks of aging. Cell. 153: 1194–1217.

Luo, J., A.Y. Nikolaev, S. Imai, D. Chen, F. Su, A. Shiloh et al. 2001. Negative control of p53 by Sir2alpha promotes cell survival under stress. Cell. 107: 137–148.

McDonald, R.B. and R.C. Ruhe. 2011. Aging and longevity: why knowing the difference is important to nutrition research. Nutrients. 3: 274–282.

McDonnell, E., B.S. Peterson, H.M. Bomze and M.D. Hirschey. 2015. SIRT3 regulates progression and development of diseases of aging. Trends Endocrinol. Metab. 26: 486–492.

Meyer, T.E., S.J. Kovacs, A.A. Ehsani, S. Klein, J.O. Holloszy and L. Fontana. 2006. Long-term caloric restriction ameliorates the decline in diastolic function in humans. J. Am. Coll. Cardiol. 47: 398–402.

Motta, M.C., N. Divecha, M. Lemieux, C. Kamel, D. Chen, W. Gu et al. 2004. Mammalian SIRT1 represses forkhead transcription factors. Cell. 116: 551–563.

Nicholas, L.M., J.L. Morrison, L. Rattanatray, S. Zhang, S.E. Ozanne and I.C. McMillen. 2015. The early origins of obesity and insulin resistance: timing, programming and mechanisms. Int. J. Obes. (Lond). 40: 229–238.

Notterman, D.A. and C. Mitchell. 2015. Epigenetics and understanding the impact of social determinants of health. Pediatr. Clin. North. Am. 62: 1227–1240.

Nunn, A.V., G.W. Guy and J.D. Bell. 2014. The intelligence paradox; will ET get the metabolic syndrome? Lessons from and for Earth. Nutr. Metab. (Lond). 11: 34.

Oberdoerffer, P., S. Michan, M. McVay, R. Mostoslavsky, J. Vann, S.K. Park et al. 2008. SIRT1 redistribution on chromatin promotes genomic stability but alters gene expression during aging. Cell. 135: 907–918.

Oellerich, M.F. and M. Potente. 2012. FOXOs and sirtuins in vascular growth, maintenance, and aging. Circ. Res. 110: 1238–1251.

Partridge, L. 2001. Evolutionary theories of ageing applied to long-lived organisms. Exp. Gerontol. 36: 641–650.

Ravussin, E., L.M. Redman, J. Rochon, S.K. Das, L. Fontana, W.E. Kraus et al. 2015. A 2-year randomized controlled trial of human caloric restriction: feasibility and effects on predictors of health span and longevity. J. Gerontol. A Biol. Sci. Med. Sci. 70: 1097–1104.

Redman, L.M., W.E. Kraus, M. Bhapkar, S.K. Das, S.B. Racette, C.K. Martin et al. 2014. Energy requirements in nonobese men and women: results from CALERIE. Am. J. Clin. Nutr. 99: 71–78.

Reusens, B., S.E. Ozanne and C. Remacle. 2007. Fetal determinants of type 2 diabetes. Curr. Drug Targets. 8: 935–941.

Rizza, W., N. Veronese and L. Fontana. 2014. What are the roles of calorie restriction and diet quality in promoting healthy longevity? Ageing Res. Rev. 13: 38–45.

Ruetenik, A. and A. Barrientos. 2015. Dietary restriction, mitochondrial function and aging: from yeast to humans. Biochim. Biophys. Acta. 1847: 1434–1447.

Schilling, M.M., J.K. Oeser, J.N. Boustead, B.P. Flemming and R.M. O'Brien. 2006. Gluconeogenesis: re-evaluating the FOXO1-PGC-1alpha connection. Nature. 443: E10–11.

Sengupta, S., T.R. Peterson, M. Laplante, S. Oh and D.M. Sabatini. 2010. mTORC1 controls fasting-induced ketogenesis and its modulation by ageing. Nature. 468: 1100–1104.

Shinmura, K., K. Tamaki, K. Saito, Y. Nakano, T. Tobe and R. Bolli. 2007. Cardioprotective effects of short-term caloric restriction are mediated by adiponectin via activation of AMP-activated protein kinase. Circulation. 116: 2809–2817.

Schones, D.E., A. Leung and R. Natarajan. 2015. Chromatin modifications associated with diabetes and obesity. Arterioscler. Thromb. Vasc. Biol. 35: 1557–1561.

Sedivy, J.M., G. Banumathy and P.D. Adams. 2008. Aging by epigenetics—a consequence of chromatin damage? Exp. Cell Res. 314: 1909–1917.

Stein, P.K., A. Soare, T.E. Meyer, R. Cangemi, J.O. Holloszy and L. Fontana. 2012. Caloric restriction may reverse age-related autonomic decline in humans. Aging Cell. 11: 644–650.

Toperoff, G., D. Aran, J.D. Kark, M. Rosenberg, T. Dubnikov, B. Nissan et al. 2012. Genome-wide survey reveals predisposing diabetes type 2-related DNA methylation variations in human peripheral blood. Hum. Mol. Genet. 21: 371–383.

Villareal, D.T., L. Fontana, S.K. Das, L. Redman, S.R. Smith, E. Saltzman et al. 2016. Effect of two-year caloric restriction on bone metabolism and bone mineral density in non-obese younger adults: a randomized clinical trial. J. Bone Miner. Res. 31: 40–51.

Winnik, S., J. Auwerx, D.A. Sinclair and C.M. Matter. 2015. Protective effects of sirtuins in cardiovascular diseases: from bench to bedside. Eur. Heart J. 36: 3404–3412.

Zampieri, M., F. Ciccarone, R. Calabrese, C. Franceschi, A. Burkle and P. Caiafa. 2015. Reconfiguration of DNA methylation in aging. Mech. Ageing Dev. 151: 60–70.

Zhu, M., G.D. Lee, L. Ding, J. Hu, G. Qiu, R. de Cabo et al. 2007. Adipogenic signaling in rat white adipose tissue: modulation by aging and calorie restriction. Exp. Gerontol. 42: 733–744.

Chapter 9

Role of Dietary Pattern and Obesity on Telomere Homeostasis

Sonia García-Calzón[1,2] and *Amelia Marti del Moral*[2,*]

INTRODUCTION

An important link has been found between telomere deregulation and several age-related diseases, such as cancer and metabolic diseases (Sanders and Newman 2013, Rizvi et al. 2014). Short telomeres have been described in peripheral blood cells from patients with cancer, type 2 diabetes, metabolic syndrome and in obese individuals.

Notably, environmental factors are involved in the development of obesity and other age-related disease. It is reported that unhealthy diets could increase metabolic diseases risk and reduced longevity (Crous-Bou et al. 2014). One mechanism by which diet may influence health outcomes is through its potential effects on the length of telomeres, the repeating DNA–protein structures that protect the ends of linear chromosomes and helps to maintain genome stability (Bethancourt et al. 2015).

In this chapter, firstly we examine the role of dietary intake in telomere homeostasis specifically addressing the impact of a Mediterranean dietary pattern. Secondly, we summarize the available literature on the relationship between TL and obesity elucidating the potential mechanisms involved in this important connection.

Genetic and Lifestyle Influences on Telomere Length

TL is strongly influenced by genetic factors with heritability estimates ranging from 36 to 84% (Aviv 2012). Several genetic variants are shown to be involved in telomere

[1] Epigenetics and Diabetes Unit, Lund University Diabetes Centre, Department of Clinical Sciences, CRC, Scania University Hospital, 205 02 Malmö, Sweden.

[2] Department of Nutrition, Food Sciences and Physiology, School of Pharmacy and Nutrition, University of Navarra, Pamplona, Spain.

* Corresponding author: amarti@unav.es

maintenance in Genome-wide association studies (GWAS) (Codd et al. 2013, Lee et al. 2013). Specifically, common variants related to *TERC* and *TERT* locus were observed to be associated with TL and longevity in GWAS studies (Bojesen et al. 2013, Codd et al. 2013). Moreover, in a meta-analysis with 37,684 individuals seven loci associated with leukocyte TL were found (Codd et al. 2013). Most of the loci are linked to telomere biology pathways (*TERC*, *TERT*, *NAF1*, *OBFC1* and *RTEL1*).

However, genetic factors do not fully explain TL, and it is necessary to take into consideration the effect of modifiable factors. Moreover, there is substantial variability in the rate of telomere shortening independent of chronological age suggesting that telomere erosion is likely a modifiable factor (Aviv et al. 2009). There is also evidence that lifestyle factors may modify the health and lifespan of an individual by changing TL. In this sense, a number of studies show a relationship between healthy lifestyle and longer TL. Sun et al. (Sun et al. 2012) observed that an adherence to a healthy lifestyle, defined by 5 modifiable factors (smoking, physical activity, adiposity, alcohol intake and diet), was associated with longer TL in U.S. women. Another study found the same relationship in U.S. men but defining the healthy lifestyle by 4 low-risk factors (low or no smoking, higher intake of fruit and vegetables, lower BMI and higher levels of physical activity) (Mirabello et al. 2009).

Interestingly, results from the Nurses' Health Study (United States) showed that adherence to the Mediterranean diet together with an overall healthy lifestyle did delay telomere shortening (Towsend et al. 2016). Meanwhile, high adiposity (larger BMI or waist circumference) was able to promote telomere attrition.

Dietary Pattern and Telomere Length

Food choice can significantly affect our telomeres, health, and longevity (Rafie et al. 2016). Recently, Freitas-Simoes et al. (2016) summarizes the evidence about dietary influences (nutrients, foods, and dietary patterns) on TL in an excellent review paper.

Here we will focus on the influence of food item related to the Mediterranean diet on TL. We have summarized the findings from human studies in Table 1.

High intake of fruit, vegetables (Diaz et al. 2010, Tiainen et al. 2012) or dietary fiber (Cassidy et al. 2010) has been associated with longer telomeres. On the contrary, a high consumption of processed meat (Nettleton et al. 2008) or alcohol (Strandberg et al. 2012, Sun et al. 2012) was associated with telomere shortening. Furthermore, an inverse association between total fat or specific fatty acids, such as polyunsaturated fatty acids (PUFA), monounsaturated fatty acids (MUFA) (Cassidy et al. 2010, Kark et al. 2012) and saturated fatty acids (Tiainen et al. 2012, Song et al. 2013) with TL was reported. And also it appears that a higher intake in omega-3 fatty acids is beneficial for slowing down the rate of telomere shortening (Farzaneh-Far et al. 2010, Kiecolt-Glaser et al. 2013, O'Callaghan et al. 2014).

The "Mediterranean diet" was first defined by Keys and Grande (1957), and it refers to a food pattern that provides a characteristic profile of nutrient intake with a high MUFA:SFA (monounsaturated fatty acids:saturated fatty acids) ratio due to a high intake of olive oil; together with high intakes of fiber, vitamins, folate, natural

Table 1. Effect of the Mediterranean diet and food components on telomere length.

	Characteristics	Duration/Cell type	Subjects	Main results	Author
MeDiet	MeDiet enriched in MUFA by virgin olive oil	4 week intervention -endothelial cells	20 elderly subjects	Prevent TS	Marin et al. 2012
	Better adherence to a traditional MeDiet	Cross-sectional -leukocyte	217 elderly subjects	Associated to lower rate of TS & higher telomerase activity	Boccardi et al. 2013
	Greater adherence to the MeDiet	Cross-sectional -leukocyte	4676 disease-free women (Nurses' Health Study)	Associated with LT	Crous-Bou et al. 2014
	Higher adherence to the MeDiet	Cross-sectional Multi-ethnic -leukocyte	residents of New York aged 65 years or older	Associated with LT among whites	Gu et al. 2015
	Higher adherence to the MeDiet pattern	Longitudinal -leukocyte	Ala 12 carriers of PPARG gene from PREDIMED-NAVARRA	Prevention of TS after 5 years	Garcia-Calzón et al. 2015
	Higher baseline adherence to the MeDiet	Cross-sectional -leukocyte	520 subjects from PREDIMED-NAVARRA	Associated with LT in women	Garcia-Calzón et al. 2016
Fruits and vegetables	In women fruit and vegetable consumption	Cross-sectional -leukocyte	318 subjects free of disease	Attenuated the shorter TL and coronary artery calcium relation	Diaz et al. 2010
	Fruit and vegetable consumption	Cross-sectional -leukocyte	1942 men and women from the Helsinki Birth Cohort Study	Positively associated with TL in women	Tiainen et al. 2012
Dietary fiber	Dietary fiber intake	Cross-sectional -leukocyte	2284 female participants from the Nurses' Health Study	Positively associated with TL (z score)	Cassidy et al. 2010
Processed meat	Higher processed meat intake	Cross-sectional -leukocyte	840 adults from the Multi-Ethnic Study of Atherosclerosis	Associated with lower TL	Netleton et al. 2010

Table 1 contd. ...

...Table 1 contd.

Table 1. Effect of the Mediterranean diet and food components on telomere length.

	Characteristics	Duration/Cell type	Subjects	Main results	Author
Alcohol	Minor alcohol consumption in midlife	Cross-sectional -leukocyte	499 men (mean age 76 years	Associated with shorter TL in old age	Strandberg et al. 2012
	Alcohol in moderation (1 drink/week to 2 drinks/day)	Cross-sectional -leukocyte	5,862 women within the NHS cohort	Associated with longer TL	Sun et al. 2010
Omega 3	Baseline blood levels of marine omega-3 fatty acids	Follow-up -leukocyte	608 ambulatory outpatients with stable coronary artery disease	Inverse relationship with the rate of TS over 5 years	Farzaneh-Far 2010
	4 months of n-3 PUFA supplementation, RCT	Follow-up -leukocyte	106 healthy sedentary overweight	TL increased with decreasing n-6:n-3 ratio	Kiecolt-Glaser 2013
	6 months of n-3 PUFA	Follow-up -leukocyte	33 adults ages > 65 y with cognitive impairment	Higher DHA levels associated with lower TS	O'Callaghan 2014
MUFA	MUFA intake	Longitudinal -leukocyte	609 subjects	Inversely related to LTL	Kark et al. 2012
PUFA	PUFA intake specifically linoleic acid	Cross-sectional -leukocyte	2284 female participants from the Nurses' Health Study	Inversely associated with TL	Cassidy 2010
SFA	Total fat and SFA intake	Cross-sectional -leukocyte	1942 men and women from the Helsinki Birth Cohort Study	Inversely associated with TL in men	Tiainen et al. 2012
	Intake of short-to-medium-chain SFA	Cross-sectional -leukocyte	4029 healthy women	Associated with shorter TL	Song et al. 2013

TS, telomere shortening; LT, longer telomeres

antioxidants, and a low intake of animal protein (Galbete et al. 2015). It is postulated to be beneficial against cardiovascular disease, being olive oil consumption, the main source of fat, with important cardio-protective effect. Also, tree nuts, rich in polyunsaturated fatty acids, are an integral part of the Mediterranean food pattern. Previous studies found a protective association between nut consumption and risk of cardiovascular disease, diabetes, or weight gain (Razquin et al. 2009, Martinez-Gonzalez and Sanchez-Villegas 2004, Psaltopoulou et al. 2004).

Although MeDiet has been widely considered a model of healthy eating, there are few studies on the potential beneficial role of a Mediterranean dietary pattern on telomere integrity (Willett et al. 1995). There are several studies exploring the health benefits of the MeDiet on aging and a longer lifespan, and also associated with a better cardiovascular risk profile (Trichopoulou et al. 2003, Buckland et al. 2011, Esposito et al. 2011, Estruch et al. 2013). In fact, several follow-up studies indicated that the adherence to a typical MeDiet is associated with lower mortality and increased longevity (Trichopoulou and Dilis 2007, Roman et al. 2008, Fernández del Río et al. 2016). The health effects of a high consumption of olive oil, especially extra virgin olive oil (EVOO), seem to be linked to antioxidant, anti-inflammatory and anti-microbial activities (Martin-Pelaez et al. 2013).

The evidence concerning the role of adherence to a MeDiet pattern on telomere integrity is also compiled in Table 1. In elderly subjects Marin et al. (Marin et al. 2012) observed that a 4-week intervention with a MeDiet (enriched in virgin olive oil) prevents telomere shortening. Meanwhile, Boccardi et al. (Boccardi et al. 2013) reported a decrease in telomere shortening rate and an increase in peripheral blood mononuclear cells telomerase activity in elderly subjects consuming traditional MeDiet. Notably, Gu et al. (2015) in an elderly multi-ethnic population of New York City found that a higher adherence to a MeDiet was associated with longer leukocytes TL among whites but not among African Americans and Hispanics.

In line with this, in the frame of the Nurses' Health Study with 4676 women, Crous-Bou et al. (Crous-Bou et al. 2014) reported that a greater adherence to the MeDiet was associated with longer telomeres (Table 1). Moreover, our lab has recently shown that a greater baseline adherence to a Mediterranean dietary pattern was associated with longer telomeres also in women at high cardiovascular risk belonging to the PREDIMED-Navarra trial (Garcia-Calzon 2015, Garcia-Calzon et al. 2016). Interestingly, we also found an apparent gene-diet interaction through the observed changes in the MUFA+PUFA/carbohydrates ratio, as this ratio increased telomere lengthening was higher in the Ala carriers of the PPARG gene compared with the Pro/Pro subjects after 5 years of a MeDiet intervention (Garcia-Calzon et al. 2015a). All these findings suggest the benefits of adherence to the MeDiet for promoting health and longevity.

In recent years, analysis of dietary patterns has been an approach used in nutritional epidemiology which allows the measurement of the human diet as a whole and the assessment of the combined effect of the diet. Specifically, dietary indexes are one of the methodological tools for a holistic assessment of diet. In this direction, our research group was the first to report both cross-sectional and longitudinal associations between the inflammatory potential of the diet and telomere shortening in subjects with a high cardiovascular disease risk (García-Calzón et al. 2015b). In addition, our findings are consistent with a beneficial effect of adherence to an anti-inflammatory diet on aging and health by slowing down telomere shortening. Moreover, we also show that longer telomeres were associated with higher dietary total antioxidant capacity and lower white bread consumption in Spanish children and adolescents (Garcia-Calzon et al. 2015c).

These findings might open a new line of investigation about the potential role of an antioxidant and anti-inflammatory diet in TL homeostasis, and their mechanisms implicated.

Obesity and Telomere Length

Telomere length and obesity development: potential mechanisms

Obesity is a multifactorial chronic disease characterized by the hypertrophy and hyperplasia of the adipose tissue, as a result of a positive balance between energy intake and energy expenditure (Chatzigeorgiou et al. 2014). Obesity, in particular excess visceral adiposity, is associated with metabolic imbalances such as insulin resistance, hyperglycaemia, dyslipidaemia and hypertension (Esser et al. 2014). In fact, these metabolic disorders increase the risk of development of type 2 diabetes and cardiovascular disease contributing to high rates of mortality and morbidity (Esser et al. 2014). Thus, more studies are needed to understand this complex disease that has reached epidemic proportions globally, with at least 2.8 million people dying each year as a result of being overweight or obese (WHO 2016).

Obesity is characterized by a low-level but chronic inflammatory state. Excess adiposity increases the production of a wide range of adipokines, including hormones, cytokines and immune factors that presented pro-inflammatory actions (Ouchi et al. 2011). Moreover, adipose tissue in obese individuals is infiltrated by a large number of macrophages, and this recruitment is linked to systemic inflammation and insulin resistance (Weisberg et al. 2003). Obesity per se may also induce systemic oxidative stress and this could be the underlying cause of dysregulation of adipokines and development of metabolic disorders (Furukawa et al. 2004). Thus, this imbalance in body composition is now recognized as a state of increased oxidative stress and inflammation for the organism (Furukawa et al. 2004). Therefore, these two mechanisms are responsible for increasing the risk of the development of obesity-related diseases.

Telomeres get shorter as people get older, leading to consider TL as a biomarker for biological aging and also for age-related diseases (Blackburn 2010). But it is well-known that telomere attrition is likely to be a modifiable factor as there is substantial variability in the rate of telomere shortening that is independent of chronological age (Aviv 2009, Garcia-Calzon 2015). In this sense, chronic oxidative stress and inflammation have been reported as the main underlying mechanisms responsible for telomere shortening (Aviv 2009, Epel 2012). Indeed, telomeres are highly sensitive to the detrimental action of hydroxyl radicals due to their high content of guanines (Kawanishi and Oikawa 2004) and inflammation also shorten TL through negative regulation of telomerase activity (Beyne-Rauzy et al. 2005). Considering that obesity accelerates aging and is accompanied by oxidative stress and inflammation, it could be these mechanisms the ones that link this chronic disease with telomere shortening.

Actually, chronic inflammation and oxidative stress linked to obesity have been related to accelerated telomere shortening.

Association between adiposity indexes and telomere length

The most common method of measuring obesity is the Body Mass Index (BMI). Several studies have assessed the association between BMI and TL to better understand whether telomere shortening is related to obesity (Table 2).

Quite a few studies reported a negative association between BMI and leukocyte TL (Valdes et al. 2005, Nordfjall et al. 2008, Lee et al. 2011, Cui et al. 2013, Rode et al. 2014, Muezzinler et al. 2016), whereas others did not find a correlation with this parameter (Diaz et al. 2010, Zhu et al. 2011). Noteworthy, Rode et al. (Rode et al. 2014) studied leukocyte TL in 45,069 individuals from the Copenhagen General Population Study, showing that leukocyte TL decreased in 7 bp per unit of increase in BMI. Accordingly, another large work also showed that increased BMI was associated with shorter leukocyte TL ($P = 7 \times 10^{-14}$) in 4,576 individuals (Weischer et al. 2014). A more recent study performed in 3,600 older adults reported that BMI and TL associations varied according to age because BMI was inversely associated with TL in only those participants younger than 60 years (–6 bp per unit of increase in BMI), and not in the elderly (Muezzinler et al. 2016). Furthermore, a meta-analysis provided a systematic review of studies on the relationship between BMI and leukocyte TL (Muezzinler et al. 2014). For cross-sectional studies, the pooled estimates for correlation and regression coefficients were –0.057 (95% CI: –0.102 to –0.012) and $-0.008\,kBP \cdot kg \cdot m^{-2}$ (95% CI: –0.016 to 0.000), respectively, suggesting a biologically plausible inverse association between BMI and leukocyte TL in adults (Muezzinler et al. 2014).

Not only BMI has been associated with leukocyte telomere length, but also other anthropometric parameters (Buxton et al. 2014) such as waist circumference (Nordfjall et al. 2008, Al-Attas et al. 2010, Cui et al. 2013, Garcia-Calzon et al. 2014a) or waist to height ratio (WHtR) (Cui et al. 2013, Garcia-Calzon et al. 2014a). Interestingly, a study carried out in 3,256 American Indians (14–93 years old, 60% women) found that leukocyte TL was significantly and inversely associated with all of the studied obesity parameters, including BMI, waist circumference, WHtR, waist to hip ratio and percentage of body fat (Chen et al. 2014). Since these measurements are considered indices of central obesity and are associated with insulin resistance, shorter telomeres may also be associated with abdominal obesity increasing the morbidity risk.

Case-control studies: telomere length and obesity risk

Case-control studies do not demonstrate causation though are very efficient to identify an association between an exposure and an outcome, in this case between obese subjects and TL. In 2005, Valdes et al. (Valdes et al. 2005) first reported that telomeres of obese women were 240 bp shorter than those of lean women.

Table 2. Studies assessing the association between telomere length and obesity or adiposity.

Main findings	Type of study	Sample characteristics	Publication
Children and adolescents			
Obese children have a mean LTL that is 23.9% shorter than that of their non-obese counterparts	Case-control study	793 French children aged 2–17 years	Buxton et al. 2011
Adiposity and adipokines are not associated with LTL	Cross-sectional	667 adolescents (aged 14–18 years)	Zhu et al. 2011
Obese boys, but not girls, present shorter LTL than the non-obese ones	Cross-sectional	148 Arab children (5–12 years)	Al-Attas et al. 2010
No difference in LTL between obese and non-obese participants	Cross-sectional	53 Italian children (age 8.2 ± 3.5 years)	Zannolli et al. 2008
Weight loss intervention is accompanied by an increase in LTL. Initial longer LTL predicts a better weight loss response	Longitudinal	74 obese adolescents aged 12–16 years	García-Calzón et al. 2014
Adults			
Telomeres of obese women are 240 bp shorter than those of lean women	Cross-sectional	1,122 white women	Valdes et al. 2005
Inverse association between LTL and body anthropometric indices	Cross-sectional	2,912 Chinese women	Cui et al. 2013
LTL is associated with an obesity phenotype but only in women	Cross-sectional	989 individuals from the Malmö Diet and Cancer Cohort and the Northern Sweden MONICA project	Nordfjall et al. 2008
Higher total and abdominal adiposity is associated with shorter LTL	Cross-sectional	309 white participants from The Fels Longitudinal Study	Lee et al. 2011
Obese participants have shorter LTL than non-obese individuals	Cross-sectional	3,256 American Indians participating in the Strong Heart Family Study	Chen et al. 2014
Increased body weight is associated with shorter LTL cross-sectionally, but not with LTL change after 10 years	Cross-sectional and longitudinal	4,576 Danish people	Weischer et al. 2014
High BMI is associated with short LTL	Cross-sectional	45,069 individuals from the Copenhagen General Population Study	Rode et al. 2014
Inverse association between LTL and BMI	Cross-sectional	193 middle-age Arabs	Al-Attas et al. 2010
No linear associations between LTL and several measures of obesity	Cross-sectional	317 subjects	Díaz et al. 2010

Table 2 contd. ...

...Table 2 contd.

Table 2. Studies assessing the association between telomere length and obesity or adiposity.

Main findings	Type of study	Sample characteristics	Publication
Subcutaneous adipose tissue cells from obese subjects show shortened telomeres	Cross-sectional	72 subjects	Moreno-Navarrete et al. 2010
Telomere length of adipocytes correlates negatively with waist and adipocyte size	Cross-sectional	21 obese women	el Bouazzaoui et al. 2014
Obese subjects have shorter telomeres than non-obese individuals	Cross-sectional	23 Caucasian subjects	Zanolli et al. 2008
LTL is inversely associated with multiple measures of adiposity. BMI increase in women is associated with shorter telomeres later in life	Cross-sectional and longitudinal	5,598 participants	Buxton et al. 2014
Positive association between body weight status and LTL after 17 years of follow-up	Longitudinal	2,749 individuals from the Health and Retirement Study	An and Yan 2016
Obese subjects have shorter telomeres compared to non-obese subjects. LTL shows an additional attrition after bariatric surgery	Case-control and weight-loss bariatric surgery	237 subjects	Formichi et al. 2014
Increases in relative LTL are found after 10 years of bariatric surgery	Weight-loss bariatric surgery	142 obese subjects	Laimer et al. 2015
Increased telomere length was observed after 12 and 52 weeks of a calorie-restricted diet in midrectal biopsies	Weight-loss intervention	12 obese men	O'Callaghan et al. 2009
No change in LTL after 12 months of weight-loss and/or exercise intervention program	Randomized weight-loss clinical trial	439 postmenopausal women	Mason et al. 2013
Weight gain is associated with accelerated telomere attrition after 10 years follow-up	Longitudinal	70 individuals	Gardner et al. 2005
Inverse association of LTL and self-reported weight gain and cycling	Cross-sectional and longitudinal	647 women	Kim et al. 2009
LTL at baseline is associated with positive change in BMI and body fat after 7 years	Longitudinal	2,721 elderly subjects (42% black and 58% white)	Njajou et al. 2012
Basal LTL is inversely associated with changes in obesity parameters after 5 years of a Mediterranean Diet intervention. An increase in LTL is observed when obesity indices improve	Longitudinal	521 subjects from the PREDIMED study	García-Calzón et al. 2014a
Telomere lengthening after weight-loss bariatric surgery intervention	Weight-loss intervention	37 subjects	Carulli et al. 2016

Table 2 contd. ...

...Table 2 contd.

Table 2. Studies assessing the association between telomere length and obesity or adiposity.

Main findings	Type of study	Sample characteristics	Publication
High maternal adiposity during pregnancy and higher weight gain during the first year of life are associated with shorter telomeres in older women but not in men	Longitudinal	1082 subjects from the Helsinki Birth Cohort Study	Guzzardi et al. 2016
Weight gain during adulthood and obesity may contribute to shorter LTL, below 60 years of age	Cross-sectional and longitudinal	3,600 older adults	Müezzinler et al. 2016
Higher baseline waist circumference and glucose, and lower HDL are associated with shorter LTL after 6 years follow-up	Longitudinal	1,808 participants	Révész et al. 2015
A biologically plausible inverse association is observed between BMI and LTL	Meta-analysis	16 studies	Muezzinler et al. 2014
Thirty-nine studies show either weak or moderate correlation between obesity and LTL; but they show an important heterogeneity	Meta-analysis	63 studies comprising 119,439 subjects	Mundstock et al. 2016

LTL: leukocyte telomere length

Interestingly, a large study published in 2014 including 3,256 American Indians found that obese participants had significantly shorter leukocyte TL than non-obese individuals (P = 0.0002) (Chen et al. 2014). Similar results were observe elsewhere (Formichi et al. 2014). Furthermore, another study reported that obese adults have shorter telomeres than their normal-weight counterparts, but they did not observed this phenomenon in children (Zannolli et al. 2008). On the other hand, a case-control study showed obese children to have a mean leukocyte TL of 23.9% shorter than that of their non-obese counterparts (Buxton et al. 2011). Likely, Al-Attas et al. (Al-Attas et al. 2010) observed that obese boys presented shorter leukocyte TL than the non-obese ones, but this result was not evident in girls.

The most common way to assess the association between obesity and TL in human studies is measuring telomeres in leukocytes, since blood is the most accessible and non-invasive tissue. But analysing TL in target tissues for disease, such as adipose tissue, could shed light about the underlying mechanisms implicated in obesity. Four studies so far have measured TL of adipose tissue in lean and obese subjects (Moreno-Navarrete et al. 2010, Monickaraj et al. 2012, el Bouazzaoui et al. 2014, Lakowa et al. 2015). All these studies came to the same conclusion that the association of shorter telomeres with obesity disease is not only happening in leukocytes but also in adipose tissue. Notably, BMI could contribute independently to TL variance in subcutaneous and visceral adipose tissue (Moreno-Navarrete et al. 2010).

Longitudinal studies: impact of obesity on telomere shortening

Few studies in the literature have analyzed the influence of changes in adiposity on telomere shortening after a follow-up period. Weight-loss interventions suggested to be efficient in slowing down TL (O'Callaghan et al. 2009, Garcia-Calzon et al. 2014b). For instance, within the EVASYON study, an increased leukocyte TL was found after a 2-month weight loss intensive program, modifying diet and physical activity, in obese adolescents (Garcia-Calzon et al. 2014b). On the contrary, a 12-month randomized intervention trial of weight loss and/or exercise was not sufficient to yield significant changes in leukocyte TL in postmenopausal women (Mason et al. 2013). In this line, TL increased after weight loss induced by bariatric surgery within 10 years after surgery (Laimer et al. 2015), whereas it has been observed that TL was not restored after 1 year of bariatric surgery (Formichi et al. 2014).

Weight gain has been associated with accelerated telomere attrition in several studies. It has been found that increasing abdominal adiposity is accompanied by accelerated telomere attrition (Garcia-Calzon et al. 2014a, Revesz et al. 2015). Remarkably, Gardner et al. (Gardner et al. 2005) observed that weight gain was associated with accelerated telomere attrition in 70 young adults after 10 years of follow-up. In agreement, Kim et al. (Kim et al. 2009) found an inverse association of TL and self-reported weight gain and cycling in 647 women. However, other studies reported no longitudinal relationship between TL and adiposity (Hovatta et al. 2012, Weischer et al. 2014).

Telomere length as a biomarker for adiposity

Mostly, all of the literature considered obesity as a risk factor for telomere shortening. But the opposite could also be true since the causal relationship of obesity and TL remains open. Indeed, some studies have elucidated a potential role of TL in predicting changes in adiposity markers. A prospective study by Njajou et al. (Njajou et al. 2012) reported leukocyte TL at baseline to be associated with changes in BMI and percentage of body fat after 7-years of follow-up in an elderly population. In agreement, findings from the PREDIMED study revealed that a higher baseline leukocyte TL was associated with a better improvement in anthropometric parameters, including central adiposity, after 5 years of a nutritional intervention (Garcia-Calzon et al. 2014a). There is just one study reporting that initial longer telomeres could be a potential predictor for a better weight loss response after 2 months of an intervention program in adolescents (Garcia-Calzon et al. 2014b). A larger study, including 2,848 participants age 18–65 years, further supports these findings since they found that shorter baseline TL was associated with unfavorable scores of most metabolic syndrome components at the 2 and 6-year follow-up (Revesz et al. 2014). In this context, the assessment of TL could be a useful biomarker for a better understanding of the biological pathways implicated in obesity progression. However, more controlled longitudinal larger studies are needed to investigate this issue, and to shed light on whether telomere shortening is a cause or a consequence of changes in adiposity.

In summary, obesity, one of the most prevalent health concerns in developed and undeveloped countries, is also an age-related disease. Telomeres could act as markers for adiposity and help in the design and management of lifestyle interventions including Mediterranean dietary pattern aimed to promote health and reduce age-related diseases.

Keywords: Telomere length, obesity, Mediterranean diet, longitudinal studies, aged-related diseases

References

Al-Attas, O.S., N. Al-Daghri, A. Bamakhramah, S. Shaun Sabico, P. McTernan and T.T. Huang. 2010. Telomere length in relation to insulin resistance, inflammation and obesity among Arab youth. Acta Paediatr. 99: 896–899.

An, R. and H. Yan. 2016. Body weight status and telomere length in U.S. middle-aged and older adults. Obes. Res. Clin. Pract. (in press).

Aviv, A. 2009. Leukocyte telomere length: the telomere tale continues. Am. J. Clin. Nutr. 89: 1721–1722.

Aviv, A. 2012. Genetics of leukocyte telomere length and its role in atherosclerosis. Mutat. Res. 730: 68–74.

Beyne-Rauzy, O., N. Prade-Houdellier, C. Demur, C. Recher, J. Ayel, G. Laurent et al. 2005. Tumor necrosis factor-alpha inhibits hTERT gene expression in human myeloid normal and leukemic cells. Blood. 106: 3200–3205.

Bethancourt, H.J., M. Kratz, S.A. Beresford, M.G. Hayes, C.W. Kuzawa, P.L. Duazo et al. 2015. No association between blood telomere length and longitudinally assessed diet or adiposity in a young adult Filipino population. Eur. J. Nutr. (in press).

Blackburn, E.H. 2010. Telomeres and telomerase: the means to the end (Nobel lecture). Angew. Chem. Int. Ed. Engl. 49: 7405–7421.

Bojesen, S.E., K.A. Pooley, S.E. Johnatty, J. Beesley, K. Michailidou, J.P. Tyrer et al. 2013. Multiple independent variants at the TERT locus are associated with telomere length and risks of breast and ovarian cancer. Nat. Genet. 45: 371–384, 384e371–372.

Buckland, G., A. Agudo, N. Travier, J.M. Huerta, L. Cirera, M.J. Tormo et al. 2011. Adherence to the mediterranean diet reduces mortality in the Spanish cohort of the European Prospective Investigation into Cancer and Nutrition (EPIC-Spain). Br. J. Nutr. 106: 1581–1591.

Buxton, J.L., R.G. Walters, S. Visvikis-Siest, D. Meyre, P. Froguel and A.I. Blakemore. 2011. Childhood obesity is associated with shorter leukocyte telomere length. J. Clin. Endocrinol. Metab. 96: 1500–1505.

Buxton, J.L., S. Das, A. Rodriguez, M. Kaakinen, A. Couto Alves, S. Sebert et al. 2014. Multiple measures of adiposity are associated with mean leukocyte telomere length in the northern Finland birth cohort 1966. PLoS One. 9: e99133.

Carulli, L., C. Anzivino, E. Baldelli, M.F. Zenobii, M.B. Rocchi and M. Bertolotti. 2016. Telomere length elongation after weight loss intervention in obese adults. Mol. Genet. Metab. 118: 138–42.

Cassidy, A., I. De Vivo, Y. Liu, J. Han, J. Prescott, D.J. Hunter et al. 2010. Associations between diet, lifestyle factors, and telomere length in women. Am. J. Clin. Nutr. 91: 1273–1280.

Chatzigeorgiou, A., E. Kandaraki, A.G. Papavassilioua and M. Koutsilieris. 2014. Peripheral targets in obesity treatment: a comprehensive update. Obes. Rev. 15: 487–503.

Chen, S., F. Yeh, J. Lin, T. Matsuguchi, E. Blackburn, E.T. Lee et al. 2014. Short leukocyte telomere length is associated with obesity in American Indians: the Strong Heart Family study. Aging (Albany NY). 6: 380–389.

Codd, V., C.P. Nelson, E. Albrecht, M. Mangino, J. Deelen, J.L. Buxton et al. 2013. Identification of seven loci affecting mean telomere length and their association with disease. Nat. Genet. 45: 422–427, 427e421–422.

Crous-Bou, M., T.T. Fung, J. Prescott, B. Julin, M. Du, Q. Sun et al. 2014. Mediterranean diet and telomere length in Nurses' Health Study: population based cohort study. BMJ. 349: g6674.

Cui, Y., Y.T. Gao, Q. Cai, S. Qu, H. Cai, H.L. Li et al. 2013. Associations of leukocyte telomere length with body anthropometric indices and weight change in Chinese women. Obesity (Silver Spring). 21: 2582–2588.

Diaz, V.A., A.G. Mainous, M.S. Player and C.J. Everett. 2010. Telomere length and adiposity in a racially diverse sample. Int. J. Obes. (Lond.). 34: 261–265.

El Bouazzaoui, F., P. Henneman, P. Thijssen, A. Visser, F. Koning, M.A. Lips et al. 2014. Adipocyte telomere length associates negatively with adipocyte size, whereas adipose tissue telomere length associates negatively with the extent of fibrosis in severely obese women. Int. J. Obes. (Lond.). 38: 746–749.

Epel, E. 2012. How "reversible" is telomeric aging? Cancer Prev. Res. (Phila). 5: 1163–1168.

Esposito, K., C. Di Palo, M.I. Maiorino, M. Petrizzo, G. Bellastella, I. Siniscalchi et al. 2011. Long-term effect of mediterranean-style diet and calorie restriction on biomarkers of longevity and oxidative stress in overweight men. Cardiol. Res. Pract. 2011: 293916.

Esser, N., S. Legrand-Poels, J. Piette, A.J. Scheen and N. Paquot. 2014. Inflammation as a link between obesity, metabolic syndrome and type 2 diabetes. Diabetes Res. Clin. Pract. 105: 141–150.

Estruch, R., E. Ros, J. Salas-Salvado, M.I. Covas, D. Corella, F. Aros et al. 2013. Primary prevention of cardiovascular disease with a Mediterranean diet. N. Engl. J. Med. 368: 1279–1290.

Farzaneh-Far, R., J. Lin, E.S. Epel, W.S. Harris, E.H. Blackburn and M.A. Whooley. 2010. Association of marine omega-3 fatty acid levels with telomeric aging in patients with coronary heart disease. JAMA. 303: 250–257.

Fernández del Río, L., E. Gutiérrez-Casado, A. Varela-López and J.M. Villalba. 2016. Olive Oil and the Hallmarks of Aging. Molecules. 21: 163.

Formichi, C., S. Cantara, C. Ciuoli, O. Neri, F. Chiofalo, F. Selmi et al. 2014. Weight loss associated with bariatric surgery does not restore short telomere length of severe obese patients after 1 year. Obes. Surg. 24: 2089–2093.

Freitas-Simoes, T.M., E. Ros and A. Sala-Vila. 2016. Nutrients, foods, dietary patterns and telomere length: Update of epidemiological studies and randomized trials. Metabolism. 65: 406–15.

Furukawa, S., T. Fujita, M. Shimabukuro, M. Iwaki, Y. Yamada, Y. Nakajima et al. 2004. Increased oxidative stress in obesity and its impact on metabolic syndrome. J. Clin. Invest. 114: 1752–1761.

Galbete, C., E. Toledo, J.B. Toledo, M. Bes-Rastrollo, P. Buil-Cosiales, A. Marti et al. 2015. Mediterranean diet and cognitive function: the SUN project. J. Nutr. Health Aging. 19: 305–12.

Garcia-Calzon, S. 2015. Telomere Length in Different Spanish Age Groups: Association with Diet, Genetics and Adiposity traits. Ph.D. dissertation. University of Navarra, Pamplona, Spain.

Garcia-Calzon, S., A. Gea, C. Razquin, D. Corella, R.M. Lamuela-Raventos, J.A. Martinez et al. 2014a. Longitudinal association of telomere length and obesity indices in an intervention study with a Mediterranean diet: the PREDIMED-NAVARRA trial. Int. J. Obes. (Lond.). 38: 177–182.

Garcia-Calzon, S., A. Moleres, A. Marcos, C. Campoy, L.A. Moreno, M.C. Azcona-Sanjulian et al. 2014b. Telomere length as a biomarker for adiposity changes after a multidisciplinary intervention in overweight/obese adolescents: the EVASYON study. PLoS One. 9: e89828.

García-Calzón, S., M.A. Martínez-González, C. Razquin, D. Corella, J. Salas-Salvadó, J.A. Martínez et al. 2015a. Pro12Ala polymorphism of the PPARγ2 gene interacts with a mediterranean diet to prevent telomere shortening in the PREDIMED-NAVARRA randomized trial. Circ. Cardiovasc. Genet. 8: 91–9.

García-Calzón, S., G. Zalba, M. Ruiz-Canela, N. Shivappa, J.R. Hébert, J.A. Martínez et al. 2015b. Dietary inflammatory index and telomere length in subjects with a high cardiovascular disease risk from the PREDIMED-NAVARRA study: cross-sectional and longitudinal analyses over 5 y. Am. J. Clin. Nutr. 102: 897–904.

García-Calzón, S., A. Moleres, M.A. Martínez-González, J.A. Martínez, G. Zalba, A. Marti; GENOI members. 2015c. Dietary total antioxidant capacity is associated with leukocyte telomere length in a children and adolescent population. Clin. Nutr. 34: 694–9.

García-Calzón, S., M.A. Martínez-González, C. Razquin, F. Arós, J. Lapetra, J.A. Martínez et al. 2016. Mediterranean diet and telomere length in high cardiovascular risk subjects from the PREDIMED-NAVARRA study. Clin. Nutr. 35: 1399–1405.

Gardner, J.P., S. Li, S.R. Srinivasan, W. Chen, M. Kimura, X. Lu et al. 2005. Rise in insulin resistance is associated with escalated telomere attrition. Circulation. 111: 2171–2177.

Gu, Y., L.S. Honig, N. Schupf, J.H. Lee, J.A. Luchsinger, Y. Stern et al. 2015. Mediterranean diet and leukocyte telomere length in a multi-ethnic elderly population. Age (Dordr.). 37: 24.

Guzzardi, M.A., P. Iozzo, M.K. Salonen, E. Kajantie and J.G. Eriksson. 2016. Maternal adiposity and infancy growth predict later telomere length: a longitudinal cohort study. Int. J. Obes. 40: 1063–9.

Hovatta, I., V.D. de Mello, L. Kananen, J. Lindstrom, J.G. Eriksson, P. Ilanne-Parikka et al. 2012. Leukocyte telomere length in the finnish diabetes prevention study. PLoS One. 7: e34948.

Kark, J.D., N. Goldberger, M. Kimura, R. Sinnreich and A. Aviv. 2012. Energy intake and leukocyte telomere length in young adults. Am. J. Clin. Nutr. 95: 479–487.

Kawanishi, S. and S. Oikawa. 2004. Mechanism of telomere shortening by oxidative stress. Ann. N. Y. Acad. Sci. 1019: 278–284.

Keys, A. and F. Grande. 1957. Role of dietary fat in human nutrition. III. Diet and the epidemiology of coronary heart disease. Am. J. Public Health Nations Health. 47: 1520–1530.

Kiecolt-Glaser, J.K., E.S. Epel, M.A. Belury, R. Andridge, J. Lin, R. Glaser et al. 2013. Omega-3 fatty acids, oxidative stress, and leukocyte telomere length: A randomized controlled trial. Brain Behav. Immun. 28: 16–24.

Kim, S., C.G. Parks, L.A. DeRoo, H. Chen, J.A. Taylor, R.M. Cawthon et al. 2009. Obesity and weight gain in adulthood and telomere length. Cancer Epidemiol. Biomarkers Prev. 18: 816–820.

Laimer, M., A. Melmer, C. Lamina, J. Raschenberger, P. Adamovski, J. Engl et al. 2015. Telomere length increase after weight loss induced by bariatric surgery: results from a 10 years prospective study. Int. J. Obes. (Lond.). 40: 773–778.

Lakowa, N., N. Trieu, G. Flehmig, T. Lohmann, M.R. Schon, A. Dietrich et al. 2015. Telomere length differences between subcutaneous and visceral adipose tissue in humans. Biochem. Biophys. Res. Commun. 457: 426–432.

Lee, M., H. Martin, M.A. Firpo and E.W. Demerath. 2011. Inverse association between adiposity and telomere length: The fels longitudinal study. Am. J. Hum. Biol. 23: 100–106.

Lee, J.H., R. Cheng, L.S. Honig, M. Feitosa, C.M. Kammerer, M.S. Kang et al. 2013. Genome wide association and linkage analyses identified three loci-4q25, 17q23.2, and 10q11.21-associated with variation in leukocyte telomere length: the long life family study. Front Genet. 4: 310.

Marin, C., J. Delgado-Lista, R. Ramirez, J. Carracedo, J. Caballero, P. Perez-Martinez et al. 2012. Mediterranean diet reduces senescence-associated stress in endothelial cells. Age (Dordr.). 34: 1309–1316.

Martin-Pelaez, S., M.I. Covas, M. Fito, A. Kusar and I. Pravst. 2013. Health effects of olive oil polyphenols: recent advances and possibilities for the use of health claims. Mol. Nutr. Food Res. 57: 760–771.

Martínez-González, M.A. and A. Sánchez-Villegas. 2004. The emerging role of Mediterranean diets in cardiovascular epidemiology: monounsaturated fats, olive oil, red wine or the whole pattern? Eur. J. Epidemiol. 19: n9–13.

Mason, C., R.A. Risques, L. Xiao, C.R. Duggan, I. Imayama, K.L. Campbell et al. 2013. Independent and combined effects of dietary weight loss and exercise on leukocyte telomere length in postmenopausal women. Obesity (Silver Spring). 21: E549–554.

Mirabello, L., W.Y. Huang, J.Y. Wong, N. Chatterjee, D. Reding, E.D. Crawford et al. 2009. The association between leukocyte telomere length and cigarette smoking, dietary and physical variables, and risk of prostate cancer. Aging Cell. 8: 405–413.

Monickaraj, F., K. Gokulakrishnan, P. Prabu, C. Sathishkumar, R.M. Anjana, J.S. Rajkumar et al. 2012. Convergence of adipocyte hypertrophy, telomere shortening and hypoadiponectinemia in obese subjects and in patients with type 2 diabetes. Clin. Biochem. 45: 1432–1438.

Moreno-Navarrete, J.M., F. Ortega, M. Sabater, W. Ricart and J.M. Fernandez-Real. 2010. Telomere length of subcutaneous adipose tissue cells is shorter in obese and formerly obese subjects. Int. J. Obes. (Lond.). 34: 1345–1348.

Muezzinler, A., A.K. Zaineddin and H. Brenner. 2014. Body mass index and leukocyte telomere length in adults: a systematic review and meta-analysis. Obes. Rev. 15: 192–201.

Muezzinler, A., U. Mons, A.K. Dieffenbach, K. Butterbach, K.U. Saum, M. Schick et al. 2016. Body mass index and leukocyte telomere length dynamics among older adults: Results from the ESTHER cohort. Exp. Gerontol. 74: 1–8.

Nettleton, J.A., A. Diez-Roux, N.S. Jenny, A.L. Fitzpatrick and D.R. Jr. Jacobs. 2008. Dietary patterns, food groups, and telomere length in the Multi-Ethnic Study of Atherosclerosis (MESA). Am. J. Clin. Nutr. 88: 1405–1412.

Njajou, O.T., R.M. Cawthon, E.H. Blackburn, T.B. Harris, R. Li, J.L. Sanders et al. 2012. Shorter telomeres are associated with obesity and weight gain in the elderly. Int. J. Obes. (Lond.). 36: 1176–1179.

Nordfjall, K., M. Eliasson, B. Stegmayr, O. Melander, P. Nilsson and G. Roos. 2008. Telomere length is associated with obesity parameters but with a gender difference. Obesity (Silver Spring). 16: 2682–2689.

O'Callaghan, N.J., P.M. Clifton, M. Noakes and M. Fenech. 2009. Weight loss in obese men is associated with increased telomere length and decreased abasic sites in rectal mucosa. Rejuvenation Res. 12: 169–176.

O'Callaghan, N., N. Parletta, C.M. Milte, B. Benassi-Evans, M. Fenech and P.R. Howe. 2014. Telomere shortening in elderly individuals with mild cognitive impairment may be attenuated with omega-3 fatty acid supplementation: a randomized controlled pilot study. Nutrition. 30: 489491.

Ouchi, N., J.L. Parker, J.J. Lugus and K. Walsh. 2011. Adipokines in inflammation and metabolic disease. Nat. Rev. Immunol. 11: 85–97.

Psaltopoulou, T., A. Naska, P. Orfanos, D. Trichopoulos, T. Mountokalakis and A. Trichopoulou. 2004. Olive oil, the Mediterranean diet, and arterial blood pressure: the Greek European Prospective Investigation into Cancer and Nutrition (EPIC) study. Am. J. Clin. Nutr. 80: 1012–1018.

Rafie, N., S. Golpour Hamedani, F. Barak, S.M. Safavi and M. Miraghajani. 2016. Dietary patterns, food groups and telomere length: a systematic review of current studies. Eur. J. Clin. Nutr. (in press).

Razquin, C., J.A. Martinez, M.A. Martinez-Gonzalez, D. Corella, J.M. Santos and A. Marti. 2009. The Mediterranean diet protects against waist circumference enlargement in 12Ala carriers for the PPARgamma gene: 2 years' follow-up of 774 subjects at high cardiovascular risk. Br. J. Nutr. 102: 672–679.

Revesz, D., Y. Milaneschi, J.E. Verhoeven and B.W. Penninx. 2014. Telomere length as a marker of cellular aging is associated with prevalence and progression of metabolic syndrome. J. Clin. Endocrinol. Metab. 99: 4607–4615.

Revesz, D., Y. Milaneschi, J.E. Verhoeven, J. Lin and B.W. Penninx. 2015. Longitudinal associations between metabolic syndrome components and telomere shortening. J. Clin. Endocrinol. Metab. 100: 3050–3059.

Rizvi, S., S.T. Raza and F. Mahdi. 2014. Telomere length variations in aging and age-related diseases. Curr. Aging Sci. 7: 161–167.

Rode, L., B.G. Nordestgaard, M. Weischer and S.E. Bojesen. 2014. Increased body mass index, elevated C-reactive protein, and short telomere length. J. Clin. Endocrinol. Metab. 99: E1671–1675.

Roman, B., L. Carta, M.A. Martinez-Gonzalez and L. Serra-Majem. 2008. Effectiveness of the Mediterranean diet in the elderly. Clin. Interv. Aging. 3: 97–109.

Sanders, J.L. and A.B. Newman. 2013. Telomere length in epidemiology: a biomarker of aging, age-related disease, both, or neither? Epidemiol. Rev. 35: 112–31.

Song, Y., N.C. You, M.K. Kang, L. Hou, R. Wallace, C.B. Eaton et al. 2013. Intake of small-to-medium chain saturated fatty acids is associated with peripheral leukocyte telomere length in postmenopausal women. J. Nutr. 143: 907–914.

Strandberg, T.E., A.Y. Strandberg, O. Saijonmaa, R.S. Tilvis, K.H. Pitkala and F. Fyhrquist. 2012. Association between alcohol consumption in healthy midlife and telomere length in older men. The Helsinki Businessmen Study. Eur. J. Epidemiol. 27: 815–822.

Sun, Q., L. Shi, J. Prescott, S.E. Chiuve, F.B. Hu, I. De Vivo et al. 2012. Healthy lifestyle and leukocyte telomere length in U.S. women. PLoS One. 7: e38374.

Tiainen, A.M., S. Mannisto, P.A. Blomstedt, E. Moltchanova, M.M. Perala, N.E. Kaartinen et al. 2012. Leukocyte telomere length and its relation to food and nutrient intake in an elderly population. Eur. J. Clin. Nutr. 66: 1290–1294.

Townsend, M.K., H. Aschard, I. De Vivo, K.B. Michels and P. Kraft. 2016. Genomics, telomere length, epigenetics, and metabolomics in the nurses' health studies. Am. J. Public Health. 106: 1663–1668.

Trichopoulou, A., T. Costacou, C. Bamia and D. Trichopoulos. 2003. Adherence to a Mediterranean diet and survival in a Greek population. N. Engl. J. Med. 348: 2599–2608.

Trichopoulou, A. and V. Dilis. 2007. Olive oil and longevity. Mol. Nutr. Food Res. 51: 1275–1278.

Valdes, A.M., T. Andrew, J.P. Gardner, M. Kimura, E. Oelsner, L.F. Cherkas et al. 2005. Obesity, cigarette smoking, and telomere length in women. Lancet. 366: 662–664.

Weisberg, S.P., D. McCann, M. Desai, M. Rosenbaum, R.L. Leibel and A.W. Jr. Ferrante. 2003. Obesity is associated with macrophage accumulation in adipose tissue. J. Clin. Invest. 112: 1796–1808.

Weischer, M., S.E. Bojesen and B.G. Nordestgaard. 2014. Telomere shortening unrelated to smoking, body weight, physical activity, and alcohol intake: 4,576 general population individuals with repeat measurements 10 years apart. PLoS Genet. 10: e1004191.

Willett, W.C., F. Sacks, A. Trichopoulou, G. Drescher, A. Ferro-Luzzi, E. Helsing et al. 1995. Mediterranean diet pyramid: a cultural model for healthy eating. Am. J. Clin. Nutr. 61: 1402S–1406S.

Zannolli, R., A. Mohn, S. Buoni, A. Pietrobelli, M. Messina, F. Chiarelli et al. 2008. Telomere length and obesity. Acta Paediatr. 97: 952–954.

Zhu, H., X. Wang, B. Gutin, C.L. Davis, D. Keeton, J. Thomas et al. 2011. Leukocyte telomere length in healthy Caucasian and African-American adolescents: relationships with race, sex, adiposity, adipokines, and physical activity. J. Pediatr. 158: 215–220.

Chapter 10

Advanced Research in Telomeres and Disease Risk

Virginia Boccardi, Irene Tassi* and *Patrizia Mecocci*

INTRODUCTION

Telomeres are specialized structures localized at the very ends of human chromosomes whose primary function is to prevent a cell from sensing linear chromosome ends as breaks in the DNA. Telomeres are composed of a long sequence of TTAGGG, which together with specialized proteins, form a cap like structure, suppressing the activation of DNA damage response (DDR). They were first described in 1938 and further investigation led to the discovery of an intracellular enzyme which was responsible for telomere elongation. This enzyme works as a reverse transcriptase and it is called "telomerase." Both telomere and telomerase are required for preserving genome integrity and stability. Due to the "end replication problem" and telomerase repression, in most normal human somatic cells, telomeres progressively shorten with each cell division and replication event, leading to replicative cellular senescence. Telomere shortening has been proposed to function as a "mitotic clock" that measures how many times a cell has divided (Sfeir et al. 2005, De Lange 2002), and it can be accelerated by many factors such as inflammation and oxidative stress (Zhang et al. 2015).

Telomere length has been considered a very attractive marker of replicative history of somatic cells: it provides a quantitative measurement related to biological mechanisms such as proliferation and senescence; it appears to record the number of past cell divisions; it is reproducible in the individual and highly correlated across tissues; it is associated with adverse lifestyle and other risk factors, as well as with age, race and gender.

Institute of Gerontology and Geriatrics, Department of Medicine, University of Perugia, Italy.
* Corresponding author: virginia.boccardi@unipg.it

However, whether telomere length may be considered as an effective biomarker of aging is still unclear and under debate (Sanders and Newman 2013), even if shorter telomeres, as measured in leukocyte (Leukocyte Telomere Length: LTL), have been consistently found in many age-related diseases (Fig. 1). In fact, numerous observational studies demonstrate that short LTL or abnormalities in genetics regulation of telomeres length are associated with hypertension (Demissie et al. 2006), type 2 diabetes mellitus (Murillo-Ortiz et al. 2012), cardiovascular disorders (Saliques et al. 2010), neurodegenerative diseases and cancer. Some findings, however, have also been contradictory or even paradoxical, thus cross-sectional data on measurements of telomere length are obviously not enough to evaluate the full importance of time-dependent changes associated with disease risk. In this chapter, we will discuss about the more recent finding in telomeres biology and risk of the most prevalent age related diseases.

Type 2 Diabetes

There are growing body of evidence supporting the association between shorter LTL and type 2 diabetes mellitus (T2D) (Elks and Scott 2014). However, most available studies have a cross-sectional design, which precludes a causal inference. T2D is mainly caused by a combination of peripheral insulin resistance and β-cell dysfunction associated with impaired β-cell regeneration and reduced β-cell mass. In mice, deletion of telomerase gene and its subsequent down-expression is associated with a significative reduction in replication capacity of β-cell, which leads to a reduced islet mass and failure to produce adequate amounts of insulin in response to glucose stimulation or high fat diet (Kuhlow et al. 2010). It has been also proposed that the accumulation of short telomeres in pancreatic tissue can compromise β-cell signaling and survival, reducing insulin secretion ability through inhibition of calcium-mediated exocytosis (Guo et al. 2011). Whether telomere shortening contributes to diabetes pathogenesis or it is a consequence of the disease is still

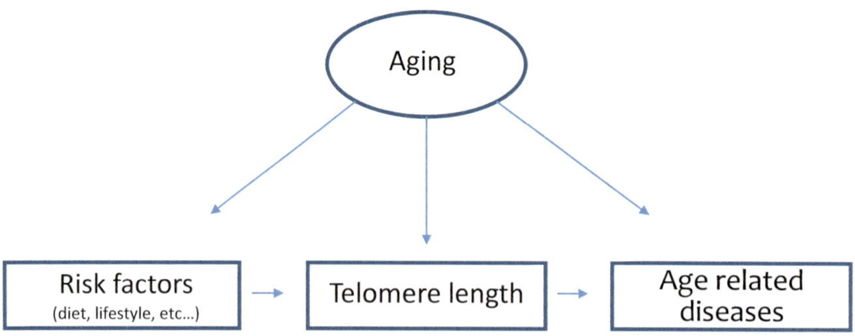

Fig. 1. Conceptual framework relating telomere length and age-related diseases risk.

unclear and under debate. Shorter telomeres may lead to premature β-cell senescence, resulting in reduced β-cell mass and subsequent impaired insulin secretion and glucose tolerance. Conversely, elevated blood glucose levels significantly increase oxidative stress and inflammation, potentially interfering with telomere erosion rate and telomerase function, resulting in shortened telomeres, both in pancreatic tissue and leukocytes. In fact, chronic hyperglycemia, advanced glycation products (AGEs) accumulation, inflammation and oxidative stress, underlying such a disease, can significantly accelerate LTL erosion rates. The high guanine content of telomeres makes them particularly vulnerable to reactive oxygen species (Houben et al. 2008), which is significantly increased in plasma of subjects affected by T2D (Roberts and Sindhu 2009). It has been also demonstrated that chronic hyperglycemia is able to promote cytokines secretion (such as IL1β) that exacerbates the shortening of telomeres by inducing a proinflammatory condition and increased cell turnover (Salpea et al. 2013). From all cross sectional data available, it is still unclear whether this attrition is actually the cause or the resultant consequence of diabetes, however, considering that it seems to be associated with disease progression, LTL can be a useful marker for T2D risk. Interestingly, Zhao and colleagues (Zhao et al. 2013) recently investigated the association of LTL at baseline with future risk of diabetes over an average follow-up period of five years. They found that shorter LTL is associated with future development of T2D independently of known type 2 diabetes risk factors. The authors also found that individuals in the lowest quartile of LTL were at almost twice the risk of developing T2D compared with those with longer telomeres. Indeed, a positive correlation has been demonstrated between telomere attrition rate and increase in the number of diabetic complications (Fuller et al. 1990), such as diabetic nephropathy (Verzola et al. 2008) and microalbuminuria (Tentolouris et al. 2007).

Hypertension

Hypertension, one of the most common age related disorder, is associated with both insulin resistance and oxidative stress (Saad et al. 2004, Sowers and Frohlich 2004), which significantly affect telomere erosion rate; thus it has been hypothized that telomere dynamics may be also a risk marker for hypertensions. According to this hypothesis, it has been shown that telomere length is inversely related with pulse pressure such that individuals having a relatively narrow pulse pressure have longer LTL (Jeanclos et al. 2000). Later, Yang and colleagues (2009) demonstrated that LTL is significantly shorter in patients with essential hypertension than in normotensive subjects, in agreement with previous studies (Demissie et al. 2006). Some evidence suggest that telomere shortening may play also a role in the pathogenesis of hypertension (Fuster et al. 2007) probably inducing genomic instability related to an early cellular senescence, as reported by Giannotti and colleagues (2010) who described shorter LTL in progenitor endothelial cells of prehypertensive and hypertensive patients. However, in the Framingham Heart Study, hypertensive male subjects exhibited significantly shorter age-adjusted LTL compared with their normotensive peers (Demissie et al. 2006). However, the Cardiovascular Health Study did not replicate this result in a mixed male and female cohort. Thus, further studies are necessary to better clarify such an association.

Cardiovascular Diseases (CVD)

Samani and colleagues (Samani et al. 2001) who described telomere shortening in atherosclerosis first observed the association between cardiovascular diseases and shortened telomeres. Cardiovascular risk factors, such as diabetes, hypertension, male gender and cigarette's smoke, are associated with shorter LTL, even if causality remains undetermined (Fyhrquist et al. 2013). These risk factors are linked to an increase in inflammation and oxidative stress, which leads to accelerated telomere erosion, endothelial cell (EC) damage and dysfunction (Voghel et al. 2007, Nezu et al. 2015). As known, telomeres are typically shorter in men than in women and continue to shorten along with aging; in this context—in the presence of other cardiovascular risk factors—their erosion rates are significantly increased (Nilsson 2012). Benetos et al. in 2004 suggested that telomere length could be considered a strong bioindicator of cardiovascular risks (Benetos et al. 2004). In fact, in the Bruneck cohort, a prospective population-based survey on the epidemiology, pathophysiology, and prevention of cardiovascular and cerebrovascular diseases, LTL significantly correlated with family history of CVD and most importantly, baseline LTL emerged as a highly significant and independent predictor of new-onset CVD. This finding strongly suggest that the heritability of TL may contribute to the individual genetic susceptibility to CVD.

Atherosclerosis

Individuals with shorter telomeres have an increased risk of early vascular cell aging and senescence, which contributes to atherosclerosis pathogenesis. In fact, the accumulation of senescent cells, a prominent feature of atherosclerotic plaques, reduces the regenerative potential of affected tissues and promotes apoptosis, which can further exacerbate inflammatory reactions and endothelial dysfunction (Minamino et al. 2002, Matthews et al. 2006). It has also been hypothesised that, by reducing the proliferative potential of vascular smooth muscle cells, cellular senescence promotes the thinning of fibrous caps and the instability of atherosclerotic plaques (Gorenne et al. 2006). Accordingly, Wilson and colleagues (2008) demonstrated that telomere content in peripheral blood leukocytes accurately reflects that of the vascular wall, and its decrease is associated with premature vascular disease. Leukocytes of patients with atherosclerosis display remarkably shorter telomeres compared to age-matched healthy subjects (Huzen et al. 2008, Huzen et al. 2011). Interestingly, Masi and colleagues explored the association between changes in LTL over 10 years and cardiovascular phenotypes: using data from a nationally representative cohort of 60–64 years old men and women, they found that a faster rate of decrease in LTL is associated with increased in carotid intima-media thickness (IMT) (Masi et al. 2014). However, other studies did not find an association of LTL with subclinical atherosclerosis as measured by IMT (De Meyer et al. 2009). In a large, well-characterized prospective cohort of American Natives, it has been found that LTL significantly predicts incidence and progression of carotid atherosclerosis, independent of established cardiovascular risk factors including diabetes and hypertension (Chen et al. 2014). More recently, Calvert and colleagues clarified that LTL is not associated with all measures of total coronary plaque burden, such as plaque volume, rather with all markers of high-risk plaques such as culprit vessel necrotic core volume, and worse necrotic core frame plaque area (Calvert et al. 2011). In fact, in the present study, a significantly shorter LTL was found in patients with aortic dissection compared to controls. Another study confirmed that short LTL is associated with aortic dissection, even after adjustment for other risk factors (Yan et al. 2011). Thus, LTL and plaque are differently related to atherosclerotic plaque characteristics. In patients with unstable angina, inflammatory cells present in coronary artery plaques express considerably higher telomerase activity compared with circulating leukocytes (Narducci et al. 2007). Local leukocyte telomerase reactivation in plaques is thought to prolong the lifespan of inflammatory cells, so maintaining the inflammatory response. This phenomenon could be an explanation for the longer telomeres found in inflammatory lipid-rich plaques.

Coronary heart disease

Patients with history of myocardial infarction display shorter LTL, but conflicting findings have been reported on the relation between LTL and subclinical coronary artery atherosclerosis, as expressed by coronary artery calcium (CAC). Shorter LTL is associated with higher incidence of coronary heart disease (CHD) and with increased mortality (Kark et al. 2013). In a series of outpatients with stable CHD ($n = 780$), LTL has been found to be associated with all-cause mortality after a mean follow-up of 4.4 years (Farzaneh-Far et al. 2008). In an Italian prospective case-control study of adults subjects with a mean age of 40 years (Russo et al. 2012), researchers did not show neither a significant association between the LTL and acute myocardial infarction nor between LTL and the onset of major acute cardiovascular event during a medium term follow up. In contrast, Maubaret et al. (Maubaret et al. 2010) in the HIFMECH (Hypercoagulability and Impaired Fibrinolytic function Mechanisms predisposing to myocardial infarction) study previously found a significant relationship between LTL and incidence of acute myocardial infarction in older patients (60 years old), suggesting that shorter LTL could be a marker of coronary heart disease in aging. Similar results were reported by Brouilette et al. (Brouilette et al. 2008) in a case control study conducted in English patients with acute myocardial infarction at 50 years old of age and in the West of Scotland Primary Prevention Study (WOSCOPS study) (Brouilette et al. 2007) in middle aged men (45–64 years). Indeed Salpea et al. (2008), in the European Atherosclerosis Research Study II (EARSII), found shorter LTL in young male students with a paternal history of premature myocardial infarction with respect to age-matched controls with no family history of CAD. Moreover, in a large prospective study of patients with stable CAD, it has been found that LTL is associated with mortality independently of chronological age, clinical factors, echocardiographic variables and serum inflammatory markers (Farzaneh-Far et al. 2008).

Heart failure, ischaemic heart failure and stroke

Patients with heart failure have 40% reduced telomeres compared with age- and gender-matched controls, and, most importantly, the degree of telomere shortening is associated with the severity of the disease (Van der Harst et al. 2007). In a longitudinal study of 890 subjects in NYHA heart failure class II to IV, patients with shorter telomeres had an increased risk (almost twice) of death or hospitalization after 18 months of follow up. Thus, it has been hypothized that shorter telomeres might represent a risk marker in this population (Van der Harst et al. 2010). Whether this relationship represents a cause or consequence of heart failure or it is attributed to associated risk factors and coronary artery disease is uncertain (Raymond et al. 2013). However, recent evidences show that telomerase activation could be a therapeutic strategy to prevent heart failure after MI (Bär et al. 2014). Some studies suggest an overall increased risk for any stroke and ischemic stroke among individuals with shorter telomeres (Fitzpatrick et al. 2007, Fitzpatrick et al. 2011, Willeit et al. 2010, Ding et al. 2012). In contrast, a prospective study among men did not find such an association (Zee et al. 2010).

Neurological Disorders

Alzheimer's disease

Whether LTL is associated with Alzheimer's Disease (AD) is still unclear since few evidences are available and results are contradictory and inconsistent (Boccardi et al. 2015). There are studies showing shorter LTL, and other showing unchanged length in AD patients compared to control subjects. Panossian and colleagues (2003) first described that individuals with AD have shorter LTL compared to age-matched controls and that telomere length is positively correlated with cognitive performance. A study conducted in women with Down syndrome detected shorter telomeres in T cells of individuals with dementia compared to age-matched subjects without dementia (Jenkins et al. 2008). Another study (Honig et al. 2006) found reduced peripheral blood LTL in AD patients compared with control subjects. Two more recent studies suggested that AD is associated with telomere stability and functional status, rather than with telomere length (Guan et al. 2012, Guan et al. 2013). In fact, peripheral leukocytes from patients with AD were found to have similar mean telomere lengths, a decreased number of the longest telomeres and unchanged numbers of the shortest telomeres compared to controls. This result is in contrast with a previous report in cognitively healthy subjects, which showed an increased number of the shortest telomeres along with aging (Aubert and Lansdorp 2008). Other studies described similar LTL between patients with AD and healthy controls (Zekry et al. 2010, Takata et al. 2012, Moverare-Skrtic et al. 2012). Thus, the high variability of telomere lengths among studies makes its significance still unclear, suggesting that its detection cannot be considered a strong biomarker for AD risk.

Parkinson's disease

Few studies have examined the relationship between LTL and Parkinson's disease (PD). In a prospective study, Wang and colleagues (2008) found no indication that men with shorter LTL are at increased risk of PD. Rather, unexpectedly, men with shorter telomeres had a lower risk of PD then those with longer telomeres. These results have been recently confirmed (Schürks et al. 2014). In a small study on twenty-eight male subjects with PD, LTL was not different from controls. However, mean telomere length shorter than 5 kb was only found in PD patients, suggesting that telomere shortening is accelerated in PD. These apparently contradictory findings may be explained by the fact that longer telomeres in individuals with PD may reflect a higher rate of apoptosis and thus a shorter lifespan of senescent cells (which have shorter telomeres) in peripheral blood. However, this is just a hypothesis and due to the restricted number of studies and inconsistent results, further researches are needed to demonstrate that LTL may be really considered a marker of PD risk.

Cancer

Telomeres are known to preserve genome integrity and shortened telomeres lead to increased cell aging along with chromosomal instability thus increasing the risk of cancer. High levels of telomerase expression and shorter telomeres were observed in most humans cancers, whereas telomerase was found to be almost absent in most normal somatic tissues. Heterogeneity between studies may thus be due to combining related, but distinct, diseases that are influenced quite differently by LTL. A recent metanalisis demonstrated a suggestive evidence that short LTL is associated with an increase risk of cancer; the strongest evidence exists for bladder, esophageal and gastric cancer. Moreover, a longitudinal study of thirteen years in the Normative Aging Study Cohort showed that LTL decreases along with aging both in cancer-free participants and in cancer cases but more rapidly among this latter. This study strongly suggested that critical LTL might contribute to cancer initiation, which then, in turn, activates telomere maintenance mechanisms to compensate and further promote cancer, providing new insight into using LTL as an early detection biomarker of cancer (Hou et al. 2015).

Short telomeres are strong genetic susceptibility markers for bladder cancer. A study evaluating the potential association between a set of polymorphisms (SNPs) in telomere maintenance genes and bladder cancer risk identified eighteen significant SNPs: four, were from telomerase protein component 1 (TEP1) and one was from PIN2/TRF1-interacting protein 1 (PINX1) (Chang et al. 2012). Accordingly, a previous study showed that PINX1 inhibition leads to aberrant telomerase activation and telomere elongation, compromising telomere function and causing chromosomal instability, and there is evidence supporting the role of PINX1 as a tumor suppressor, acting through a telomerase-dependent mechanism (Zhou et al. 2011).

Short telomere length is significantly associated with an increased risk of esophageal squamous cell carcinoma papillomavirus-related (ESCC) (Yu et al. 2014, Xing et al. 2009), and even polymorphisms in telomere maintenance genes are associated with its increased risk (Shi et al. 2013). ESCC shows also an increased cell telomerase activity (Yu et al. 2004), even if such an activation is not related to prognosis cancer (Hsu et al. 2005). Individuals with shorter telomeres have been reported to have an increased risk for Barrett's esophageal (BE) (Risques et al. 2007). Environmental stressors that may increase oxidative damage have been associated with telomere shortening, and the increase in cancer risk itself attributable to telomere shortening appears to be potentiated by the effects of oxidative damage related to tobacco use, no use of anti-inflammatory agents and inflammation related to truncal obesity.

Gastric cancer risk is increased in people who had shorter telomeres and this risk is higher in people who had low risk for gastric cancer including H. pylori negative individuals, non-smokers, and individuals with high fruit and vegetable intake, as observed in a Polish population study (Hou et al. 2009). Comparable results have been found with a similar risk though smoking potentiated rather than minimized the risk for gastric cancer in this study population (Liu et al. 2009). Another evaluation of gastric tumors reported that telomere length was shorter in early stage cancers and then lengthened with increasing stage (Mu et al. 2012). This observation further supported that critical LTL may contribute to cancer initiation which then, in turn, activates telomere maintenance mechanisms to compensate and further promote cancer.

Conclusions

Many studies have demonstrated that LTL is associated with age-related diseases and hypotethical mechanisms are summerized in Table 1. Although rarely acknowledged, these associations do not dictate causality and for many diseases is still unclear whether LTL may be really considered a risk marker. Most human studies in telomere biology and aging related diseases are cross-sectional while largescale or well-designed studies are lacking. Moreover, LTL is not consistent and varies between individuals, organs, cell types, and most importantly even between chromosomes. Indeed there are several techniques to measure the relative and absolute telomere length in eukaryotic cells such as, Terminal Restriction Fragment (TRF), Southern Blot, Q-FISH (involves fluorescent *in situ* hybridization) and Real-Time PCR (RT-

Table 1. Hypothetical mechanisms of association between telomere length and diseases risk.

Diseases		**Mechanism/Association**
Type 2 Diabetes		• Elevated blood glucose levels and AGEs: - increase telomere erosion rate and interfere with telomere function by increasing oxidative stress - Promote pro-inflammatory cytokines secretion leading to i cell turnover • Increase telomere attrition rate leads to: - Reduction in replication capacity of β-cells - β-cells signalling impairment - Increased risk of diabetic complications
Hypertension		• Associated insulin resistance and oxidative stress affect telomere erosion rate • Telomere shortening induce genomic instability with early cellular senescence
Cardiovascular disease	• Atherosclerosis	• The accumulation of senescent cells reduces the regenerative potential and promotes apoptosis • Cellular senescence promotes the thinning of fibrous caps and the instability of plaques • Shortened telomere is associated with increased in carotid intima-media thickness and aortic dissection
	• Coronary heart disease	• Patients with shorter telomeres have higher incidence of cardiovascular diseases and increased mortality
	• Heart failure, ischaemic heart failure and stroke	• The degree of telomere shortening is associated with the severity of the disease
Neurological disorders	• Alzheimer's disease	• Few evidence available
	• Parkinson's disease	• Patients with shorter telomere have a lower risk for PD
Cancer	• Bladder cancer	• Shorter telomeres lead to increased chromosomal instability that might contribute to cancer initiation • Polymorphisms in 18 different SNPs (i.e., PINX1) can compromise telomere function and causing chromosomal instability
	• Esophageal squamous cell carcinoma	• Increased cell telomerase activity • Shorter telomeres are associated with an increased risk for Barrett's esophageal
	• Gastric cancer	• Shorter telomere are associated with increased risk of gastric cancer

PCR), which makes data not conformable. In conclusion, whether telomere biology is causally involved in the development of human diseases awaits further research from large, prospective and longitudinal as well as interventional studies.

Keywords: Telomeres, diseases, aging, telomerase, senescence

References

Aubert, G. and P.M. Lansdorp. 2008. Telomeres and aging. Physiol. Rev. 88(2): 557–79.

Bär, C., B. Bernardes de Jesus, R. Serrano, A. Tejera, E. Ayuso, V. Jimenez et al. 2014. Telomerase expression confers cardioprotection in the adult mouse heart after acute myocardial infarction. Nat. Commun. 5: 5863.

Benetos, A., J.P. Gardner, M. Zureik, C. Labat, L. Xiaobin, C. Adamopoulos et al. 2004. Short telomeres are associated with increased carotid atherosclerosis in hypertensive subjects. Hypertension. 43(2): 182–5.

Boccardi, V., L. Pelini, S. Ercolani, C. Ruggiero and P. Mecocci. 2015. From cellular senescence to Alzheimer's disease: the role of telomere shortening. Ageing Res. Rev. 22: 1–8.

Brouilette, S.W., J.S. Moore, A.D. McMahon, J.R. Thompson, I. Ford, J. Shepherd et al. 2007. Telomere length, risk of coronary heart disease, and statin treatment in the West of Scotland Primary Prevention Study: a nested case-control study. Lancet. 369: 107–114.

Brouilette, S.W., A. Whittaker, S.E. Stevens, P. Van der Harst, A.H. Goodall and N.J. Samani. 2008. Telomere length is shorter in healthy offspring of subjects with coronary artery disease: support for the telomere hypothesis. Heart. 94: 422–425.

Calvert, P.A., T. Liew, I. Gorenne, M. Clarke, C. Costopoulos, D.R. Obaid et al. 2011. Leukocyte telomere length is associated with high-risk plaques on virtual histology intravascular ultrasound and increased proinflammatory activity. Arterioscler. Thromb. Vasc. Biol. 31: 2157–2164.

Chang, J., C.P. Dinney, M. Huang, X. Wu and J. Gu. 2012. Genetic variants in telomere-maintenance genes and bladder cancer risk. Plos One. 7(2): e30665.

Chen, S., J. Lin, T. Matsuguchi, E. Blackburn, F. Yeh, L.G. Best et al. 2014. Short leukocyte telomere length predicts incidence and progression of carotid atherosclerosis in American Indians: The Strong Heart Family Study. Aging. 6(5): 414–27.

De Lange, T. 2002. Protection of mammalian telomeres. Oncogene. 21(4): 532–40.

De Meyer, T., E.R. Rietzschel, M.L. De Buyzere, M.R. Langlois, D. De Bacquer, P. Segers et al. 2009. Systemic telomere length and preclinical atherosclerosis: the Asklepios Study. Eur. Heart J. 30(24): 3074–81.

Demissie, S., D. Levy, E.J. Benjamin, L.A. Cupples, J.P. Gardner, A. Herbert et al. 2006. Insulin resistance, oxidative stress, hypertension, and leukocyte telomere length in men from the Framingham Heart Study. Aging Cell. 5: 325–330.

Ding, H., C. Chen, J.R. Shaffer, L. Liu, Y. Xu, X. Wang et al. 2012. Telomere length and risk of stroke in chinese. Stroke. 43: 658–663.

Elks, C.E. and R.A. Scott. 2014. The long and short of telomere length and diabetes. Diabetes. 63: 65–67.

Farzaneh-Far, R., R.M. Cawthon, B. Na, W.S. Browner, N.B. Schiller and M.A. Whooley. 2008. Prognostic value of leukocyte telomere length in patients with stable coronary artery disease: data from the Heart and Soul Study. Arterioscler. Thromb. Vasc. Biol. 28: 1379–1384.

Fitzpatrick, A.L., R.A. Kronmal, J.P. Gardner, B.M. Psaty, N.S. Jenny, R.P. Tracy et al. 2007. Leukocyte telomere length and cardiovascular disease in the cardiovascular health study. Am. J. Epidemiol. 165: 14–21.

Fitzpatrick, A.L., R.A. Kronmal, M. Kimura, J.P. Gardner, B.M. Psaty, N.S. Jenny et al. 2011. Leukocyte telomere length and mortality in the Cardiovascular Health Study. J. Gerontol. A Biol. Sci. Med. Sci. 66: 421–429.

Fuller, K.A., D. Pearl and C.C. Whitacre. 1990. Oral tolerance in experimental autoimmune encephalomyelitis: serum and salivary antibody responses. J. Neuroimmunol. 28(1): 15–26.

Fuster, J.J., J. Diez and V. Andrés. 2007. Telomere dysfunction in hypertension. J. Hypertens. 25(11): 2185–92.

Fyhrquist, F., O. Saijonmaa and T. Strandberg. 2013. The roles of senescence and telomere shortening in cardiovascular disease. Nat. Rev. Cardiol. 10(5): 274–83.

Giannotti, G., C. Doerries, P.S. Mocharla, M.F. Mueller, F.H. Bahlmann, T. Horvàth et al. 2010. Impaired endothelial repair capacity of early endothelial progenitor cells in prehypertension: relation to endothelial dysfunction. Hypertension. 55: 1389–1397.

Gorenne, I., M. Kavurma, S. Scott and M. Bennett. 2006. Vascular smooth muscle cell senescence in atherosclerosis. Cardiovasc. Res. 72: 9–17.

Guan, J.Z., W.P. Guan, T. Maeda and N. Makino. 2012. Effect of vitamin E administration on the elevated oxygen stress and the telomeric and subtelomeric status in Alzheimer's disease. Gerontology. 58(1): 62–9.

Guan, J.Z., W.P. Guan, T. Maeda and N. Makino. 2013. Analysis of telomere lenght and subtelomeric methylation of circulating leukocytes in women with Alzheimer's disease. Aging Clin. Exp. Res. 25(1): 17–23.

Guo, N., E.M. Parry, L.S. Li, F. Kembou, N. Lauder, M.A. Hussain et al. 2011. Short telomeres compromise β cell signaling and survival. PLoS One. 6: e17858.

Honig, L.S., N. Schupf, J.H. Lee, M.X. Tang and R. Mayeux. 2006. Shorter telomeres are associated with mortality in those with APOE epsilon4 and dementia. Ann. Neurol. 60(2): 181–7.

Hou, L., S.A. Savage, M.J. Blaser, G. Perez-Perez, M. Hoxha, L. Dioni et al. 2009. Telomere length in peripheral leukocyte DNA and gastric cancer risk. Cancer Epidemiol. Biomarkers Prev. 18(11): 3103–3109.

Hou, L., B.T. Joyce, T. Gao, L. Liu, Y. Zheng, F.J. Penedo et al. 2015. Blood telomere length attrition and cancer development in the normative aging study cohort. EBioMedicine. 2(6): 591–6.

Houben, J.M., H.J. Moonen, F.J. Van Schooten and G.J. Hageman. 2008. Telomere length assessment: biomarker of chronic oxidative stress? Free Radic. Biol. Med. 44: 235–246.

Hsu, C.P., L.W. Lee, S.E. Shai and C.Y. Chen. 2005. Clinical significance of telomerase and its associate genes expression in the maintenance of telomere length in squamous cell carcinoma of the esophagus. World J. Gastroenterol. 11(44): 6941–6947.

Huzen, J., D.J. Van Veldhuisen, W.H. Van Gilst and P. Van der Harst. 2008. Telomeres and biological ageing in cardiovascular disease. Ned. Tijdschr. Geneeskd. 152(22): 1265–70.

Huzen, J., W. Peeters, R.A. de Boer, F.L. Moll, L.S.M. Wong, V. Codd et al. 2011. Circulating leukocyte and carotid atherosclerotic plaque telomere length interrelation, association with plaque characteristics, and restenosis after endarterectomy. Arterioscler. Thromb. Vasc. Biol. 31: 1219–1225.

Jeanclos, E., N.J. Schork, K.O. Kyvik, M. Kimura, J.H. Skurnick and A. Aviv. 2000. Telomere length inversely correlates with pulse pressure and is highly familial. Hypertension. 36: 195–200.

Jenkins, E.C., L. Ye, H. Gu, S.A. Ni, C.J. Duncan, M. Velinov et al. 2008. Increased absence of telomeres may indicate Alzheimer's disease/dementia status in older individual with down syndrome. Neurosci. Lett. 440(3): 340–3.

Kark, J.D., H. Nassar, D. Shaham, R. Sinnreich, N. Goldberger, V. Aboudi et al. 2013. Leukocyte telomere length and coronary artery calcification in Palestinians. Atherosclerosis. 229(2): 363–8.

Kuhlow, D., S. Florian, G. von Figura, S. Weimer, N. Schulz, K.J. Petzke et al. 2010. Telomerase deficiency impairs glucose metabolism and insulin secretion. Aging. 2: 650–658.

Liu, X., G. Bao, T. Huo, Z. Wang, X. He and G. dong. 2009. Constitutive telomere length and gastric cancer risk: Case-control analysis in Chinese Han population. Cancer Sci. July. 100(7): 1300–1305.

Masi, S., F. D'Aiuto, C. Martin-Ruiz, T. Kahn, A. Wong, A.K. Ghosh et al. 2014. Rate of telomere shortening and cardiovascular damage: a longitudinal study in the 1946 British Birth Cohort. European Heart Journal. 35: 3296–3303.

Matthews, C., I. Gorenne, S. Scott, N. Figg, P. Kirkpatrick, A. Ritchie et al. 2006. Vascular smooth muscle cells undergo telomere-based senescence in human atherosclerosis: effects of telomerase and oxidative stress. Circ. Res. 99: 156–64.

Maubaret, C.G., K.D. Salpea, A. Jain, J.A. Cooper, A. Hamsten, J. Sanders et al. 2010. Telomeres are shorter in myocardial infarction patients compared to healthy subjects: correlation with environmental risk factors. J. Mol. Med. 88: 785–794.

Minamino, T., H. Miyauchi, T. Yoshida, Y. Ishida, H. Yoshida and I. Komuro. 2002. Endothelial cell senescence in human atherosclerosis: role of telomere in endothelial dysfunction. Circulation. 105: 1541–4.

Moverare-Skrtic, S., P. Johansson, N. Mattsson, O. Hansson, A. Wallin, J.O. Johansson et al. 2012. Leukocyte telomere length (LTL) is reduced in stable mild cognitive impairment but low LTL is not associated with conversion to Alzheimer's disease: a pilot study. Exp. Gerontol. 47(2): 179–82.

Mu, Y., Q. Zhang, L. Mei, X. Liu, W. Yang and J. Yu. 2012. Telomere shortening occurs early during gastrocarcinogenesis. Med. Oncol. 29(2): 893–898.

Murillo-Ortiz, B., F. Albarran-Tamayo, D. Arenas-Aranda, L. Benitez-Bribiesca, J.M. Malacara-Hernandez, S. Martinez-Garza et al. 2012. Telomere length and type 2 diabetes in males: a premature aging syndrome. Aging Male. 15: 54–8.

Narducci, M.L., A. Grasselli, L.M. Biasucci, A. Farsetti, A. Mule, G. Liuzzo et al. 2007. High telomerase activity in neutrophils from unstable coronary plaques. J. Am. Coll. Cardiol. 50: 2369–2374.

Nezu, T., N. Hosomi, T. Takahashi, K. Anno, S. Aoki and A. Shimamoto. 2015. Telomere G-tail length is a promising biomarker related to white matter lesions and endothelial dysfunction in patients with cardiovascular risk: A cross-sectional study. EBioMedicine. 2: 958–965.

Nilsson, P. 2012. Impact of vascular aging on cardiovascular disease: the role of telomere biology. J. Hypertens. 30 Suppl: S9–12.

Panossian, L.A., V.R. Porter, H.F. Valenzuela, X. Zhu, E. Reback, D. Mastermann et al. 2003. Telomere shortening in T cells correlates with Alzheimer's disease status. Neurobiol. Aging. 24(1): 77–84.

Raymond, A.R., G.R. Norton, P. Sareli, A.J. Woodiwiss and R.L. Brooksbank. 2013. Relationship between average leucocyte telomere length and the presence or severity of idiopathic dilated cardiomyopathy in black Africans European Journal of Heart Failure. European Journal of Heart Failure. 15(1): 54–60.

Risques, R.A., T.L. Vaughan, X. Li, R.D. Odze, P.L. Blount, K. Ayub et al. 2007. Leukocyte telomere length predicts cancer risk in Barrett's esophagus. Cancer Epidemiol. Biomarkers Prev. 16(12): 2649–2655.

Roberts, C.K. and K.K. Sindhu. 2009. Oxidative stress and metabolic syndrome. Life Sci. 84: 705–712.

Russo, A., L. Palumbo, C. Fornengo, C. Di Gaetano, F. Ricceri, S. Guarrera et al. 2012. Telomere length variation in juvenile acute myocardial infarction. Plos One. November 2012. 7(11): e49206.

Saad, M.F., M. Rewers, J. Selby, G. Howard, S. Jinagouda, S. Fahmi et al. 2004. Insulin resistance and hypertension: The insulin resistance atherosclerosis study. Hypertension. 43: 1324–31.

Saliques, S., M. Zeller, J. Lorin, L. Lorgis, J.R. Teyssier, Y. Cottin et al. 2010. Telomere length and cardiovascular disease. Arch. Cardiovasc. Dis. 103: 454–9.

Salpea, K.D., V. Nicaud, L. Tiret, P.J. Talmud and S.E. Humphries. 2008. The association of telomere length with paternal history of premature myocardial infarction in the European Atherosclerosis Research Study II. J. Mol. Med. 86: 815–824.

Salpea, K.D., C.G. Maubaret, A. Kathagen, G. Ken-Dror, D.W. Gilroy and S.E. Humphries. 2013. The effect of pro-inflammatory conditioning and/or high glucose on telomere shortening of aging fibroblasts. Plos One. September 2013. 8(9): e73756.

Samani, N.J., R. Boultby, R. Butler, J.R. Thompson and A.H. Goodall. 2001. Telomere shortening in atherosclerosis. Lancet. 358: 472–3.

Sanders, J.L. and A.B. Newman. 2013. Telomere length in epidemiology: A biomarker of aging, age-related disease, both, or neither? Epidemiol. Rev. 35: 112–31.

Schürks, M., J. Buring, R. Dushkes, J.M. Gaziano, R.Y.L. Zee and T. Kurth. 2014. Telomere length and Parkinson's disease in men: A nested case-control study 5. Eur. J. Neurol. 21(1): 93–99.

Sfeir, A.J., W. Chai, J.W. Shay and W.E. Wright. 2005. Telomere end processing: the terminal nucleotides of human chromosomes. Mol. Cell. 18: 131–138.

Shi, J., F. Sun, L. Peng, B. Li, L. Liu, C. Zhou et al. 2013. Leukocyte telomere length-related genetic variants in 1p34.2 and 14q21 loci contribute to the risk of esophageal squamous cell carcinoma. Int. J. Cancer. 132(12): 2799–2807.

Sowers, J.R. and E.D. Frohlich. 2004. Insulin and insulin resistance impact on blood pressure and cardiovascular disease. Med. Clin. North Am. 88: 63–82.

Takata, K., M. Kikukawa, H. Hanyu, Y. Takata, T. Umahara, H. Sakurai et al. 2012. Telomere length shorteming in patients with dementia with Lewy bodies. Eur. J. Neurol. 19(6): 905–10.

Tentolouris, N., R. Nzietchueng, V. Cattan, G. Poitevin, P. Lacolley, A. Papazafiropoulou et al. 2007. White blood cells telomere length is shorter in males with type 2 diabetes and microalbuminuria. Diabetes Care. 30: 2909–15.

Van der Harst, P., G. Van der Steege, R.A. de Boer, A.A. Voors, A.S. Hall, M.J. Mulder et al. 2007. Telomere length of circulating leukocytes is decreased in patients with chronic heart failure. J. Am. Coll. Cardiol. 49: 1459–1464.

Van der Harst, P., R.A. de Boer, N.J. Samani, L.S. Wong, J. Huzen, V. Codd et al. 2010. Telomere length and outcome in heart failure. Ann. Med. 42: 36–44.

Verzola, D., M.T. Gandolfo, G. Gaetani, A. Ferraris, R. Mangerini, F. Ferrario et al. 2008. Accelerated senescence in the kidneys of patients with type 2 diabetic nephropathy. Am. J. Physiol. Renal. Physiol. 295: F1563–73.

Voghel, G., N. Thorin-Trescases, N. Farhat, A. Nguyen, L. Villeneuve, A.M. Mamarbachi et al. 2007. Cellular senescence in endothelial cells from atherosclerotic patients is accelerated by oxidative stress associated with cardiovascular risk factors. Mech. Ageing Dev. 128(11-12): 662–71.

Wang, Y., Z. Hu, J. Liang, Z. Wang, J. Tang, S. Wang et al. 2008. A tandem repeat of human telomerase reverse transcriptase (hTERT) and risk of breast cancer development and metastasis in Chinese women. Carcinogenesis. 29(6): 1197–1201.

Wentzensen, I.M., L. Mirabello, R.M. Pfeiffer and S.A. Savage. 2011. The association of telomere length and cancer: a Meta-analysis. Cancer Epidemiol. Biomarkers Prev. 20(6): 1238–50.

Willeit, P., J. Willeit, A. Brandstatter, S. Ehrlenbach, A. Mayr, A. Gasperi et al. 2010. Cellular aging reflected by leukocyte telomere length predicts advanced atherosclerosis and cardiovascular disease risk. Arterioscler. Thromb. Vasc. Biol. 30: 1649–1656.

Wilson, W., K. Herbert, Y. Mistry, S. Stevens, H. Patel, R. Hastings et al. 2008. Blood leucocyte telomere DNA content predicts vascular telomere DNA content in humans with and without vascular disease. Eur. Heart J. 29(21): 2689–2694.

Xing, J., J.A. Ajani, M. Chen, J. Izzo, J. Lin, Z. Chen et al. 2009. Constitutive short telomere length of chromosome 17p and 12q but not 11q and 2p is associated with an increased risk for esophageal cancer. Cancer Prev. Res. (Phila.). 2(5): 459–465.

Yan, J., Y. Yang, C. Chen, J. Peng, H. Ding and D. Wen Wang. 2011. Short leukocyte telomere length is associated with aortic dissection. Intern. Med. 50(23): 2871–5.

Yang, Z., X. Huang, H. Jiang, Y. Zhang, H. Liu, C. Qin et al. 2009. Short telomeres and prognosis of hypertension in a chinese population. Hypertension. 53(4): 639–645.

Yu, H.P., S.Q. Xu, W.H. Lu, Y.Y. Li, F. Li, X.L. Wang et al. 2004. Telomerase activity and expression of telomerase genes in squamous dysplasia and squamous cell carcinoma of the esophagus. J. Surg. Oncol. 86(2): 99–104.

Zee, R.Y.L., P.M. Ridker and D.I. Chasman. 2011. Genetic variants of 11 telomere-pathway gene loci and therisk of incident type 2 diabetes mellitus: the women's genome health study. Atherosclerosis. 218(1): 144–146.

Zekry, D., F.R. Hermann, I. Irminger-Finger, C. Graf, C. Genet, A.M. Vitale et al. 2010. Telomere length and ApoE polymorphism in mild cognitive impairment, degenerative and vascular dementia. J. Neurol. Sci. 299(1-2): 108–11.

Zhang, J., G. Rane, X. Dai, M.K. Shanmugam, F. Arfuso, R.P. Samy et al. 2016. Ageing and the telomere connection: An intimate relationship with inflammation. Ageing Res. Rev. 25: 55–69.

Zhao, J., Y. Zhu, J. Lin, T. Matsuguchi, E. Blackburn, Y. Zhang et al. 2013. Short leukocyte telomere length predicts risk of diabetes in American Indians: the Strong Heart Family Study. Diabetes. 63: 354–362.

Zhou, X.Z., P. Huang, R. Shi, T.H. Lee, G. Lu, Z. Zhang et al. 2011. The telomerase inhibitor PinX1 is a major haploinsufficient tumor suppressor essential for chromosome stability in mice. Journal of Clinical Investigation. 121: 1266–1282.

Chapter 11

Clinical Aspects Relevant to Telomere Maintenance and Therapeutic Opportunities

Brooke R. Druliner,[1,a] *Mohamad Mouchli*[1,b] and *Lisa A. Boardman*[2,*]

INTRODUCTION

Telomeres are tandem repeated DNA sequence (TTAGGG)n that cap linear chromosomes to maintain stability (Blackburn 1991). Telomeres shorten with successive rounds of DNA replication during sequential cell division (Harley et al. 1990). Although tissues start off with equivalent telomeres, changes in telomere length occur in parallel but at different rates during the lifespan of each individual (Hastie et al. 1990, Kim et al. 2002, O'Sullivan et al. 2006). Humans begin life with roughly 13,000 base pairs of telomere in their leukocytes. In young adulthood, approximately 33 base pairs of peripheral blood leukocyte telomere length are lost annually, and this annual loss intensifies to roughly 61 base pairs of telomere after 60 years of age (Epel et al. 2004). Telomere shortening mitigates the end-replication problem, where the incomplete replication of linear DNA could result in the loss of genetic material. In healthy cells, erosion of telomere length eventually leads to regulated cell senescence and apoptosis. Telomere length and the shortening that occurs over rounds of cell division have been identified as a marker of biological age, representing a "molecular clock." Telomeres are 10–15 kilobases of gene-poor material, characterized by the

[1] Mayo Clinic, Department of Gastroenterology and Hepatology, 200 First Street SW, Rochester, MN 55905.
[a] Email: Druliner.Brooke@mayo.edu
[b] Email: Mouchli.Mohamad@mayo.edu
[2] Mayo Clinic College of Medicine, Gonda Building 9-411, 200 First Street SW, Rochester, MN 55905.
* Corresponding author: boardman.lisa@mayo.edu

binding of large protein complexes—telomere repeat binding factors (TRF1 and 2)—directly to telomeric repeats and then interact with other factors, which provides one level of telomere regulation (Blackburn 1991, 2001). This regulation broadly has a role in preventing telomere end-to-end fusions, by DNA repair mechanisms through direct interaction with TRF2 (Smogorzewska and de Lange 2004). Telomere repeat binding factors are a part of the shelterin complex, which also includes TRF1 interacting protein 2 (TIN2), protection of telomeres 1 (POT1), the POT1 and TIN2 interacting protein TPP1 and the transcriptional repressor/activator protein RAP1 (de Lange 2005a). Some of the shelterin components do not bind directly to telomere DNA, but interact and form complexes (heterodimerization) with TRF1 and TRF2. In a mouse model with the deletion of shelterin components, there were high levels of telomere dysfunction, DNA damage response and chromosomal instability (Deng et al. 2008).

Upon the initiation of the telomere-induced DNA damage response, proteins are localized to the telomere forming telomere-induced foci (TIFs) (Deng et al. 2008). Inappropriately regulated telomeres are recognized as double stranded breaks, which initiates tumor suppressor mechanisms such as cellular senescence or p53-dependent apoptosis (Rajaraman et al. 2007, Wright and Shay 1992). Therefore, telomeres themselves are now considered tumor suppressors.

Telomeres have a 3' overhang of G-rich nucleotides that is recognized by the reverse transcriptase, telomerase, one of the most common telomere maintenance mechanisms (TMM). Telomerase is encoded by the TERT gene (the reverse transcriptase component of the telomerase protein; hTERT), and extends telomeres using an RNA molecule, which is encoded by the TERC gene (hTERC or hTR) (Collins and Mitchell 2002). Telomeres can be folded into a large telomere loop, the T loop, that is caused by the G-strand overhang's ability to fold back on itself and anneal to the telomere repeats (de Lange 2004, Griffith et al. 1999). Telomerase can reverse inherent end-replication shortening of telomeres, and is active in germline or hematopoietic cells, but is typically inactive in somatic tissues. Another TMM, Alternative Lengthening of Telomeres (ALT) confers the ability to lengthen and maintain telomeres and is most commonly associated with lack of telomerase activity. ALT utilizes homologous recombination as the mechanism to extend telomeres, and is detecting commonly by the formation of loops of DNA called C-circles (Bryan et al. 1995, Henson et al. 2009).

Epigenetic modification and chromatin organization is also an important aspect of telomere structure and regulation. It has been shown that telomeres are characterized by constitutive heterochromatin, and the presence of tightly packed nucleosomes with specific histone modifications (Garcia-Cao et al. 2004). Changes in the compaction of chromatin structure at telomeres, as well as lengthening of telomeres can result in the transcriptional silencing of nearby genes in subtelomeric regions, knowns as telomere position effect (Baur et al. 2001). Since the misregulation of telomere length and maintenance mechanisms are implicated in many age-related diseases, telomere syndromes and cancer, telomeres and their maintenance mechanisms are excellent candidates for therapeutic intervention since they are critical in maintaining basic human health.

The Role of Telomeres and Telomere Maintenance Mechanisms in Disease

Telomere length changes and engagement of telomere maintenance mechanisms are implicated in many diseases, especially those that are associated with ageing (Armanios 2009). As humans age, telomeres shorten, mainly due to the fact that telomerase and other mechanisms aren't typically active in normal somatic human cells (Harley et al. 1990). Telomere shortening can lead to critically short telomeres being recognized as DNA damage and may lead to chromosomal instability, which is a contributor to age-related diseases. Markers of telomere dysfunction and DNA damage, stathmin, EF-1a, and p16 increase with age, while telomeres have a negative correlation with age (Jiang et al. 2008, Shammas 2011, Song et al. 2010). There is an association with an increased incidence of age-related diseases in people whose telomeres shorten at a rate or amount that is greater than the average for a particular age group (Cawthon et al. 2003). If telomere length changes and/or maintenance mechanisms are activated inappropriately in somatic tissues, this can lead to human diseases that are associated with age, including heart disease, Alzheimer's, ulcerative colitis, liver cirrhosis, atherosclerosis, aplastic anemia, pulmonary fibrosis, cancer, and a group of disorders known as telomere syndromes (Cawthon et al. 2003, De Lange 2005b, Oh et al. 2003, O'Sullivan et al. 2002, Roberts et al. 2014, Samani et al. 2001, Wiemann et al. 2002, Wu et al. 2003).

Diseases that have mutations (or other defects) in genes that regulate telomere maintenance, referred to as primary telomeropathies, have a broad range of overlapping symptoms and vary in the age of onset (Opresko and Shay 2016). These diseases are typically heritable and patients with mutations within the same family can exhibit different disorders or phenotypes (Opresko and Shay 2016). Although there is little known regarding the heterogeneity of these disease, the common factor is a defect in telomere maintenance among these individuals. These patients often begin life with shorter telomeres than normal individuals because they inherit impaired telomere function. The telomere-related genes that are involved in a variety of these telomeropathies are TERT, TERC, TIN2, RTEL1, CTC1, Apollo, DKC1, NHP2, NOP10, TCAB1, and PARN (Opresko and Shay 2016).

Dyskeratosis congenita—telomeropathy

A premature ageing condition known as Dyskeratosis Congenita (DC) is disease that is marked by telomere shortening that is more rapid than the normal ageing-related telomere shortening and a decrease in telomerase activity (Mitchell et al. 1999). DC patients carry mutations in genes that directly encode telomerase components or in dyskerin gene (*DKC1*) that encodes a protein involved in TERC maintenance (Vulliamy et al. 2001, Vulliamy et al. 2004). Either type of mutation results in telomere shortening, decreased telomerase activity and chromosomal instability. DC patients have an increased chance of developing spontaneous cancer, which is likely due to the chromosomal instability that occurs with this disease. Patients with mutations affecting TERT and TERC can present with aplastic anemia, prior to being diagnosed with DC or pulmonary fibrosis. Aplastic anemia has been shown to occur in conjunction with DC, but when aplastic anemia occurs in the absence of DC it is still considered a telomere syndrome because these patients frequently carry mutations in hTERT and hTERC (Yamaguchi et al. 2003, Yamaguchi et al. 2005). Similarly, 8–15% of familial idiopathic pulmonary fibrosis and 1–3% of sporadic idiopathic pulmonary fibrosis patients carry mutations in either telomerase component (Cronkhite et al. 2008, Tsakiri et al. 2007).

Telomeres and Ageing and Degenerative Diseases

Telomere length alterations have been recognized as a potential causal factor in heart disease and the degenerative Alzheimer's disease and its precursor of amnestic mild cognitive impairment (aMCI) (Boccardi et al. 2015). It is thought that a persistent DNA damage responses leading to cell senescence plays a major role in the development of Alzheimer's disease. Telomeres are implicated due to their link to the DNA damage response and senescence. It has been shown that there is both telomere shortening and lengthening in the peripheral blood leukocytes of patients with aMCI (Roberts et al. 2014). In other studies, telomere lengthening has been observed in tissues from the brains of patients with Alzheimer's (Thomas et al. 2008).

Patients with hypertension and type I and II diabetes have been shown to have short blood telomeres (Oh et al. 2003, Wong et al. 2008). Additionally, blood and endothelial cell telomeres are shorter in patients with atherosclerosis, and telomere length has been shown to accurately predict future coronary heart disease. Studies are still being done to determine if telomere length alterations are causal for these diseases and may present opportunities for intervention. In animal models, an increase in TERT expression led to increased telomerase activity and subsequent increased myocyte cell density through hyperplasia or decreased apoptosis, while in another study activation of telomerase through overexpression of TERT resulted in proliferation potential of smooth muscle cells and an extended lifespan (Minamino et al. 2001, Oh et al. 2001). Obesity is associated with an increase in reactive oxygen species as well as DNA damage, independent of age (Song et al. 2010). Studies show increases in oxidative stress and DNA damage correlated with telomere shortening in obese patients. Obese women had shorter telomere than age-matched women that were lean, and it was determined that the loss of telomeres was equivalent to the loss of 8.8 years of life, worse than the effects of smoking (Furukawa et al. 2004, Valdes et al. 2005). Other factors such as exposure to harmful agents, smoking, nature of profession and stress also impacted telomere length negatively, which has the potential to lead to age-related diseases (Shammas 2011).

Telomere, Maintenance Mechanisms and Cancer

Both telomere length and alterations in TMM are associated with many types of cancer. Telomere length is not regulated solely by attrition resulting from cell division, but also through TMM. TMM are typically inactive in somatic tissues but are often activated in cancers through exploitation of telomerase or ALT (Reddel

2003). Activation of telomerase can reverse inherent end-replication shortening of telomeres conferring an advantage for tumor cells to remain viable and proliferate, which may in part explain why rectal cancers with activation of telomerase have a worse prognosis (Tatsumoto et al. 2000, Yi et al. 2011). One study found that rectal cancers engage ALT, and also that found rectal cancers without telomerase activation (but untested for ALT) have a better prognosis than cancers with telomerase activation (Boardman et al. 2013, Yi et al. 2011).

Cellular senescence was first recognized in normal human fibroblasts, and it was noted that this process inhibited these cells from propagating indefinitely (Campisi 2000). This process, known formally as replicative senescence, involves arresting cell growth as mechanism in avoiding apoptosis, but also preventing malignant transformation. When cells have shortened telomeres and are at risk for genomic instability, they will enter senescence. However, if specific checkpoints of replicative senescence are by-passed, such as through extreme telomere shortening, chromatin modifications or mutations that cause silencing of tumor suppressors, then oncogenetic transformation occurs (Campisi 2000). In both mouse and human models, telomere dysfunction in the absence of p53 results in genomic instability and a decrease in survival (Deng et al. 2008). Mice with both the loss of p53 and short telomeres had tumors that were similar to carcinomas of the breast, skin and intestine, resembling cancer characteristic of aged humans (Liu et al. 2016). p53 is closely linked with telomere dysfunction and tumor initiation. Intact p53 and telomere dysfunction allow for tumor inhibition, whereas telomere dysfunction and defective p53 leads to tumor initiation. In fact, dysfunctional telomeres actually activate a p53-dependent senescence mechanism to suppress tumor formation (Deng et al. 2008).

The telomerase-deficient mouse has become an important model system for studying the impact of telomere and TMM alterations in an organism. Telomerase-deficient mice are generated by knocking out the Terc gene (Terc-/-); however, only a limited number of generations are possible due to infertility and diseases associated with telomere-driven ageing (Blasco et al. 1995, Blasco et al. 1997). Telomerase-deficient mice experience both a reduction of proliferative potential and apoptosis and an increase in p53 secondary to an increase of the DNA damage response (Herrera et al. 1999). Although the telomerase-deficient mouse exhibits diseases that are associated with telomere-shortening, it does not exhibit diseases with increased proliferation such as cancer indicating that further alterations are needed to produce this disease phenotype (Gonzalez-Suarez et al. 2000). In this model, short telomeres without telomerase appear to function as tumor suppressors (Gonzalez-Suarez et al. 2000). Reintroduction of the Terc gene in these mice prevents additional telomere shortening and promotes chromosomal stability and viability (Samper et al. 2001). In contrast, if Tert is overexpressed in mice adult tissues tumor development is accelerated as compared to age-matched wild-type controls and also show protection against other age-related diseases such as kidney dysfunction and infertility (Gonzalez-Suarez et al. 2001). This indicates that developing cancer for these mice requires the telomerase RNA component encoded by Terc, and modifying different components of the telomerase complex have unique effects on the pathology (Cayuela et al. 2005).

ALT is a TMM that had typically been identified as a telomerase-independent mechanism, and is dependent on homologous recombination (Cesare and Reddel 2010). Factors involved in homologous recombination (mismatch repair genes and DNA repair proteins, RAD54 and RAD51D) as well as chromatin modifiers (SUV39H histone methyltransferase and the RB family) may have a role in regulating ALT (Jeyapalan et al. 2005). However, ALT has not been described in normal processes, only in disease states or genetically modified organisms.

Dysregulation of TMM in the form of ALT is common in cancers with mesenchymal origin, and studies have shown that the activation of telomerase and ALT have distinct gene expression signatures (Lafferty-Whyte et al. 2009). ALT has been shown as a prognostic indicator in diseases; for example, in glioblastoma multiforme (Hakin-Smith et al. 2003). Following adjustment for age, the presence of ALT in these tumors conferred an increase survival time (542 days) in patients when compared to tumors without ALT (either telomerase-negative or telomerase-positive, 247 and 236 days, respectively (Hakin-Smith et al. 2003)). An interesting area of study will involve tumors with telomere length changes that are independent of both telomerase and ALT mechanisms.

Peripheral blood leukocyte telomere length may be genetically determined as a hereditary trait. Shorter telomere length in blood DNA is associated in many cancer types with higher mortality from cancer. In an evaluation of over 64,000 individuals, PBL telomere length was found to be independently associated with genetic variants in three SNPs near TERC (rs1317082); in TERT (rs7726159); and in OBFC1(rs248799) (Rode et al. 2015). Though shorter PBL telomere length is the most commonly recognized association with poorer survival and increased mortality from cancers, it has been recently determined that for cancer patients with short telomeres secondary to these genetic variants, cancer mortality is actually decreased. Cancer patients with short telomeres that were not genetically determined but instead were shorter secondary to one of the many factors that modify telomere length, such as age, sex, and psychological, physical and socioeconomic stressors, mortality from their cancer was increased. This may imply that patients with genetically determined long telomeres would have increased cancer mortality.

Patients with stages II and III colorectal cancer had a poor prognosis if they had short telomeres (Chen et al. 2014). In another study, disease free survival from colorectal cancer was shorter if the telomere length in the tumor was shorter than that of corresponding normal colon tissue (Valls et al. 2011). In up to two thirds of colorectal cancer cases, tumor telomere length has been reported to be 3–5 kb shorter than the telomeres of adjacent normal colon epithelium, which far exceeds the natural rate of telomere attrition expected in healthy tissues (Engelhardt et al. 1997, Nakamura et al. 2000). Therefore, comparisons between blood and tumor telomere length could inform survival outcomes in patients with colorectal or other cancers.

Targets and Clinical Relevance of Telomeres and Maintenance Mechanisms

Many age-related diseases are characterized by telomere shortening. This fact and studies from the telomerase-deficient mouse as well as the identification of key genes involved in telomere regulation indicate that there is potential for therapeutic intervention of these pathways. Additionally, the involvement of telomeres in cellular senescence as a tumor suppression mechanism suggests opportunities for therapeutic applications.

Re-programming stem cells to alter telomere function

The self-renewal and differentiation ability of pluripotent stem cells has made them a great source of cell therapy in humans. However, a caveat to their use is that they retain the ability to proliferate indefinitely and therefore confer a risk for tumor formation. To circumvent the issues with the therapeutic use of pluripotent stem cells, investigators have developed telomerase knockout human embryonic stem cells through gene editing (Liu et al. 2016). The knockout of telomerase does not affect the pluripotency of the stem cells, but limits their proliferative potential by controlling their telomere length. In a very important study on the use of stem cells for disease therapy, skin cells from patients with dyskeratosis congenita (DC) were reprogrammed to create induced pluripotent stem cells (iPS cells) (Agarwal et al. 2010). Patients with DC have genetic defects in the telomerase component TERC, and when diseased cells were reprogrammed they then showed upregulation of TERC (Agarwal et al. 2010). This allowed telomeres to elongate despite the original genetic defects in TERC.

Telomerase inhibitors

Since telomerase is frequently misregulated in disease, especially cancer, it is an enticing and feasible target to initiate senescence and stop proliferation of tumor cells. Telomerase inhibitors have been studied in humans (Arndt and MacKenzie 2016, Jafri et al. 2016, Mender et al. 2015). A leading telomerase inhibitor is GRN163L (Imetelstat sodium), which is a modified oligonucleotide complementary to hTERC and as a result is powerful and highly specific to telomerase (Gryaznov et al. 2007, Hochreiter et al. 2006). Through telomere shortening and induction of cellular senescence GRN163L has been shown to suppress tumor cell growth in many tested cell lines, as well as breast and lung metastases in animal models (Hochreiter et al. 2006). However, GRN163L required prolonged exposure to reduce telomere length and eventually slow tumor growth and therefore requires the additional use of neo-adjuvant therapy (Mender et al. 2015). In one study, two small-molecular inhibitors of telomerase, BIBR1532 and BIBR1591, were added to growing cells from the following cell lines: NCI-H460 lung cancer, HT1080 fibrosarcoma, MDA-MB231 breast cancer and DU145 prostate cancer cell lines. Application of these telomerase inhibitors did not affect cell viability, but exhibited progressive shortening of telomeres compared to control cells (non-cancer) (Mender et al. 2015). Using a colony-formation assay, cells treated with the telomerase inhibitor showed a 50% reduction in the formation of colonies compared to mock-treated cells. Following both transplantation into mice and treatment with the inhibitor, tumors formed at a much slower pace and prolonged treatment led to a reduced incidence of tumors without toxicity to the organism (Mender et al. 2015). In another study, 6-thio-2'-deoxyguanosine (6-thio-dG), a purine chemotherapeutic agent, was highly selective for telomerase-expression cancer cells with no toxicity at therapeutic levels (Mender et al. 2015). In a large panel of cancer cell lines, compared to normal and telomerase-negative cells, 6-thio-dG cause rapid cell death of cancer cells, but not normal cells. Additionally, 6-thio-dG caused and increased induction in TIFs and slowed the tumor growth rate after injecting mice with the A549 lung cancer cells line (Mender et al. 2015).

Agents that modify ALT and other telomere-related factors

The presence of ALT has been shown to cause chemoresistance in cancer cells (Flynn et al. 2015). Cancer cells with ALT have defects in the cell-cycle regulation of TERRA, the telomere repeat-containing RNA. The defect TERRA in ALT-positive cells is a result of ATRX loss, a chromatin remodeling protein (frequently lost in cancers). These two defects lead ultimately to the constant association of replication protein A (RPA) with telomeres following DNA replication when it is supposed to release (Flynn et al. 2015). It is thought that the persistent contact of RPA with telomeres in ALT-positive/ATRX-negative cells indicating that the protein kinase ATR is recruited by RPA as a regulator of homologous recombination and is involved in the establishment of ALT. This study showed that the ATR inhibitor VE-821 disrupted ALT and killed ALT cells *in vitro*. This inhibitor was highly selective for ALT-positive cells, which suggests that this inhibitor could be used as a specific ALT-positive cancer cell treatment (Flynn et al. 2015).

In addition to both telomerase and ALT, studies utilizing treatments targeting other cellular components implicated in telomere dysfunction and disease have shown promise (de Lange 2005a). Telomeres can form G-quadruplex structures (G4) induced by the RHPS4 ligand, which have been shown to affect normal telomere function (Salvati et al. 2007). RHPS4 causes a DNA damage response in transformed human fibroblasts and melanoma cells· which can then trigger the induction of cellular senescence. RHPS4 has also shown anti-tumor effects in mice models where human tumor cell lines were xenografted, suggesting that initiating a DNA damage response in cancer cells suppresses tumorigenesis (Salvati et al. 2007).

It has been shown the sex hormones regulate telomerase, and that both TERT expression telomerase enzymatic activity is upregulated in response to androgens in human lymphocytes and CD34+ hematopoietic cells (Bayne and Liu 2005, Calado et al. 2009, Guo et al. 2003). In a mouse model, treatment with male hormones cause telomere elongation and improved telomere dysfunction (Bar et al. 2015). In a recent study, patients with telomere diseases as a result of mutations in TERT, TERC, RTEL1 or DKC (dyskeratosis congenita patient) were treated with the synthetic male hormone with androgenic properties (danazol) (Townsley et al. 2016). Danazol resulted in telomere elongation in the circulating leukocytes, and was powered to detect a 30% improvement in telomere shortening following two years of treatment. These are another important set of findings for a powerful treatment for telomere-related diseases.

Lifestyle Interventions Affecting Telomeres

Studies have identified several life style interventions that modify telomere length. A healthy diet, exercise, social support and reduced stress decrease the pace at which blood telomeres shorten, while other dietary factors (saturated fat, processed meats) accelerate blood telomere shortening (Boccardi et al. 2013, Cherkas et al. 2008, Nettleton et al. 2008, Ornish et al. 2013, Paul 2011, Tiainen et al. 2012). Blood telomere length can be sustained or lengthened, providing opportunities for intervention. Activation of telomerase and longer telomeres in blood DNA are associated with longer survival for early stage prostate cancer patients undergoing dietary and stress management interventions (Ornish et al. 2013).

Prognostic Value of Telomeres and Relevance to Patient Care

In regards to aging and related diseases as well as poorly understood conditions, telomere length could provide a surrogate marker for biological age and age-related conditions. Understanding biological mechanisms associated with these associations and targeting genes the play a role in telomere maintenance may influence disease prevalence and progression as well as early mortality.

Development of telomere extension could improve gene therapies for some genetic diseases associated with telomere shortening (Ramunas et al. 2015). The severity of symptoms in genetic syndromes associated with telomere shortening correlates with telomere length (Vulliamy et al. 2004, Walne et al. 2008). Telomere length of peripheral blood leukocytes can predict the sustained response to some therapies such as immunosuppressive therapies, possibly serving as a biological marker for therapy failure and disease recurrence (Scheinberg et al. 2010). Rapid telomere attrition might reflect accelerated organs aging predisposing them to cancer development (Risques et al. 2008, Risques et al. 2007).

In the context of cancer, advances in molecular profiling of tumors indicate that cancer has many unique categories of disease identified by molecular distinctions. These prognostic markers are targeted at the events occurring within the tumor but are only a portion of the molecular information essential for predicting survival. Telomere length and the maintenance mechanisms in both the blood and the normal tissue from which a cancer originates may inform the patient's prognosis. Understanding the mechanism that underlies the association between shorter telomeres and diseases onset severity is critical to risk stratification and patient management.

Malignancies resulting from exposure to environmental irritants and inflammatory processes in chronic conditions are associated with regional telomere shortening (Meeker et al. 2004, O'Sullivan et al. 2002, Risques et al. 2008, Valdes et al. 2005). The effect of irritants is more pronounced in individuals who inherit short telomeres (Tsakiri et al. 2007). Studying affected individuals and families will allow changing the epidemiology of some diseases and improving diagnostic and therapeutic approaches for individuals at risk.

Knowledge about telomere-related events obtained from these studies will provide insight into individual treatment options. Studying telomere biology and maintenance mechanisms is fundamental to the understanding of pathophysiology of age-related and poorly understood conditions as well as in making diagnostic and therapeutic decisions. Understanding the role of telomeres and their maintenance in disease may amplify the meaning of biomarkers known to impact prognostication and treatment of diseases, and enhance the clinician and patient's ability to make better informed decisions regarding treatment.

Keywords: Telomeres, telomerase, ALT, telomere disease, cancer

References

Agarwal, S., Y.H. Loh, E.M. McLoughlin, J. Huang, I.H. Park, J.D. Miller et al. 2010. Telomere elongation in induced pluripotent stem cells from dyskeratosis congenita patients. Nature. 464(7286): 292–6. doi: 10.1038/nature08792.

Armanios, M. 2009. Syndromes of telomere shortening. Annu. Rev. Genomics Hum. Genet. 10: 45–61. doi: 10.1146/annurev-genom-082908-150046.

Arndt, G.M. and K.L. MacKenzie. 2016. New prospects for targeting telomerase beyond the telomere. Nat. Rev. Cancer. 16(8): 508–24. doi: 10.1038/nrc.2016.55.

Bar, C., N. Huber, F. Beier and M.A. Blasco. 2015. Therapeutic effect of androgen therapy in a mouse model of aplastic anemia produced by short telomeres. Haematologica. 100(10): 1267–74. doi: 10.3324/haematol.2015.129239.

Baur, J.A., Y. Zou, J.W. Shay and W.E. Wright. 2001. Telomere position effect in human cells. Science. 292(5524): 2075–7. doi: 10.1126/science.1062329.

Bayne, S. and J.P. Liu. 2005. Hormones and growth factors regulate telomerase activity in ageing and cancer. Mol. Cell. Endocrinol. 240(1-2): 11–22. doi: 10.1016/j.mce.2005.05.009.

Blackburn, E.H. 1991. Structure and function of telomeres. Nature. 350(6319): 569–73. doi: 10.1038/350569a0.

Blackburn, E.H. 2001. Switching and signaling at the telomere. Cell. 106(6): 661–73.

Blasco, M.A., W. Funk, B. Villeponteau and C.W. Greider. 1995. Functional characterization and developmental regulation of mouse telomerase RNA. Science. 269(5228): 1267–70.

Blasco, M.A., H.W. Lee, M.P. Hande, E. Samper, P.M. Lansdorp, R.A. DePinho et al. 1997. Telomere shortening and tumor formation by mouse cells lacking telomerase RNA. Cell. 91(1): 25–34.

Boardman, L.A., R.A. Johnson, K.B. Viker, K.A. Hafner, R.B. Jenkins, D.L. Riegert-Johnson et al. 2013. Correlation of chromosomal instability, telomere length and telomere maintenance in microsatellite stable rectal cancer: a molecular subclass of rectal cancer. PLoS One. 8(11): e80015. doi: 10.1371/journal.pone.0080015.

Boccardi, V., A. Esposito, M.R. Rizzo, R. Marfella, M. Barbieri and G. Paolisso. 2013. Mediterranean diet, telomere maintenance and health status among elderly. PLoS One. 8(4): e62781. doi: 10.1371/journal.pone.0062781.

Boccardi, V., L. Pelini, S. Ercolani, C. Ruggiero and P. Mecocci. 2015. From cellular senescence to Alzheimer's disease: The role of telomere shortening. Ageing Res. Rev. 22: 1–8. doi: 10.1016/j.arr.2015.04.003.

Bryan, T.M., A. Englezou, J. Gupta, S. Bacchetti and R.R. Reddel. 1995. Telomere elongation in immortal human cells without detectable telomerase activity. EMBO J. 14(17): 4240–8.

Calado, R.T., W.T. Yewdell, K.L. Wilkerson, J.A. Regal, S. Kajigaya, C.A. Stratakis et al. 2009. Sex hormones, acting on the TERT gene, increase telomerase activity in human primary hematopoietic cells. Blood. 114(11): 2236–43. doi: 10.1182/blood-2008-09-178871.

Campisi, J. 2000. Cancer, aging and cellular senescence. *In Vivo*. 14(1): 183–8.

Cawthon, R.M., K.R. Smith, E. O'Brien, A. Sivatchenko and R.A. Kerber. 2003. Association between telomere length in blood and mortality in people aged 60 years or older. Lancet. 361(9355): 393–5. doi: 10.1016/S0140-6736(03)12384-7.

Cayuela, M.L., J.M. Flores and M.A. Blasco. 2005. The telomerase RNA component Terc is required for the tumour-promoting effects of Tert overexpression. EMBO Rep. 6(3): 268–74. doi: 10.1038/sj.embor.7400359.

Cesare, A.J. and R.R. Reddel. 2010. Alternative lengthening of telomeres: models, mechanisms and implications. Nat. Rev. Genet. 11(5): 319–30. doi: 10.1038/nrg2763.

Chen, Y., F. Qu, X. He, G. Bao, X. Liu, S. Wan et al. 2014. Short leukocyte telomere length predicts poor prognosis and indicates altered immune functions in colorectal cancer patients. Ann. Oncol. 25(4): 869–76. doi: 10.1093/annonc/mdu016.

Cherkas, L.F., J.L. Hunkin, B.S. Kato, J.B. Richards, J.P. Gardner, G.L. Surdulescu et al. 2008. The association between physical activity in leisure time and leukocyte telomere length. Arch. Intern. Med. 168(2): 154–8. doi: 10.1001/archinternmed.2007.39.

Collins, K. and J.R. Mitchell. 2002. Telomerase in the human organism. Oncogene. 21(4): 564–79. doi: 10.1038/sj.onc.1205083.

Cronkhite, J.T., C. Xing, G. Raghu, K.M. Chin, F. Torres, R.L. Rosenblatt et al. 2008. Telomere shortening in familial and sporadic pulmonary fibrosis. Am. J. Respir. Crit. Care Med. 178(7): 729–37. doi: 10.1164/rccm.200804-550OC.

De Lange, T. 2004. T-loops and the origin of telomeres. Nat. Rev. Mol. Cell. Biol. 5(4): 323–9. doi: 10.1038/nrm1359.

De Lange, T. 2005a. Shelterin: the protein complex that shapes and safeguards human telomeres. Genes Dev. 19(18): 2100–10. doi: 10.1101/gad.1346005.

De Lange, T. 2005b. Telomere-related genome instability in cancer. Cold Spring Harb. Symp. Quant. Biol. 70: 197–204. doi: 10.1101/sqb.2005.70.032.

Deng, Y., S.S. Chan and S. Chang. 2008. Telomere dysfunction and tumour suppression: the senescence connection. Nat. Rev. Cancer. 8(6): 450–8. doi: 10.1038/nrc2393.

Engelhardt, M., P. Drullinsky, J. Guillem and M.A. Moore. 1997. Telomerase and telomere length in the development and progression of premalignant lesions to colorectal cancer. Clin. Cancer Res. 3(11): 1931–41.

Epel, E.S., E.H. Blackburn, J. Lin, F.S. Dhabhar, N.E. Adler, J.D. Morrow et al. 2004. Accelerated telomere shortening in response to life stress. Proc. Natl. Acad. Sci. USA. 101(49): 17312–5. doi: 10.1073/pnas.0407162101.

Flynn, R.L., K.E. Cox, M. Jeitany, H. Wakimoto, A.R. Bryll, N.J. Ganem et al. 2015. Alternative lengthening of telomeres renders cancer cells hypersensitive to ATR inhibitors. Science. 347(6219): 273–7. doi: 10.1126/science.1257216.

Furukawa, S., T. Fujita, M. Shimabukuro, M. Iwaki, Y. Yamada, Y. Nakajima et al. 2004. Increased oxidative stress in obesity and its impact on metabolic syndrome. J. Clin. Invest. 114(12): 1752–61. doi: 10.1172/JCI21625.

Garcia-Cao, M., R. O'Sullivan, A.H. Peters, T. Jenuwein and M.A. Blasco. 2004. Epigenetic regulation of telomere length in mammalian cells by the Suv39h1 and Suv39h2 histone methyltransferases. Nat. Genet. 36(1): 94–9. doi: 10.1038/ng1278.

Gonzalez-Suarez, E., E. Samper, J.M. Flores and M.A. Blasco. 2000. Telomerase-deficient mice with short telomeres are resistant to skin tumorigenesis. Nat. Genet. 26(1): 114–7. doi: 10.1038/79089.

Gonzalez-Suarez, E., E. Samper, A. Ramirez, J.M. Flores, J. Martin-Caballero, J.L. Jorcano et al. 2001. Increased epidermal tumors and increased skin wound healing in transgenic mice overexpressing the catalytic subunit of telomerase, mTERT, in basal keratinocytes. EMBO J. 20(11): 2619–30. doi: 10.1093/emboj/20.11.2619.

Griffith, J.D., L. Comeau, S. Rosenfield, R.M. Stansel, A. Bianchi, H. Moss et al. 1999. Mammalian telomeres end in a large duplex loop. Cell. 97(4): 503–14.

Gryaznov, S.M., S. Jackson, G. Dikmen, C. Harley, B.S. Herbert, W.E. Wright et al. 2007. Oligonucleotide conjugate GRN163L targeting human telomerase as potential anticancer and antimetastatic agent. Nucleosides Nucleotides Nucleic Acids. 26(10-12): 1577–9. doi: 10.1080/15257770701547271.

Guo, C., B.N. Armbruster, D.T. Price and C.M. Counter. 2003. *In vivo* regulation of hTERT expression and telomerase activity by androgen. J. Urol. 170(2 Pt 1): 615–8. doi: 10.1097/01.ju.0000074653.22766.c8.

Hakin-Smith, V., D.A. Jellinek, D. Levy, T. Carroll, M. Teo, W.R. Timperley et al. 2003. Alternative lengthening of telomeres and survival in patients with glioblastoma multiforme. Lancet. 361(9360): 836–8.

Harley, C.B., A.B. Futcher and C.W. Greider. 1990. Telomeres shorten during ageing of human fibroblasts. Nature. 345(6274): 458–60. doi: 10.1038/345458a0.

Hastie, N.D., M. Dempster, M.G. Dunlop, A.M. Thompson, D.K. Green and R.C. Allshire. 1990. Telomere reduction in human colorectal carcinoma and with ageing. Nature. 346(6287): 866–8. doi: 10.1038/346866a0.

Henson, J.D., Y. Cao, L.I. Huschtscha, A.C. Chang, A.Y. Au, H.A. Pickett et al. 2009. DNA C-circles are specific and quantifiable markers of alternative-lengthening-of-telomeres activity. Nat. Biotechnol. 27(12): 1181–5. doi: 10.1038/nbt.1587.

Herrera, E., E. Samper, J. Martin-Caballero, J.M. Flores, H.W. Lee and M.A. Blasco. 1999. Disease states associated with telomerase deficiency appear earlier in mice with short telomeres. EMBO J. 18(11): 2950–60. doi: 10.1093/emboj/18.11.2950.

Hochreiter, A.E., H. Xiao, E.M. Goldblatt, S.M. Gryaznov, K.D. Miller, S. Badve et al. 2006. Telomerase template antagonist GRN163L disrupts telomere maintenance, tumor growth, and metastasis of breast cancer. Clin. Cancer Res. 12(10): 3184–92. doi: 10.1158/1078-0432.CCR-05-2760.

Jafri, M.A., S.A. Ansari, M.H. Alqahtani and J.W. Shay. 2016. Roles of telomeres and telomerase in cancer, and advances in telomerase-targeted therapies. Genome Med. 8(1): 69. doi: 10.1186/s13073-016-0324-x.

Jeyapalan, J.N., H. Varley, J.L. Foxon, R.E. Pollock, A.J. Jeffreys, J.D. Henson et al. 2005. Activation of the ALT pathway for telomere maintenance can affect other sequences in the human genome. Hum. Mol. Genet. 14(13): 1785–94. doi: 10.1093/hmg/ddi185.

Jiang, H., E. Schiffer, Z. Song, J. Wang, P. Zurbig, K. Thedieck et al. 2008. Proteins induced by telomere dysfunction and DNA damage represent biomarkers of human aging and disease. Proc. Natl. Acad. Sci. USA. 105(32): 11299–304. doi: 10.1073/pnas.0801457105.

Kim, H.R., Y.J. Kim, H.J. Kim, S.K. Kim and J.H. Lee. 2002. Telomere length changes in colorectal cancers and polyps. J. Korean Med. Sci. 17(3): 360–5. doi: 10.3346/jkms.2002.17.3.360.

Lafferty-Whyte, K., C.J. Cairney, M.B. Will, N. Serakinci, M.G. Daidone, N. Zaffaroni et al. 2009. A gene expression signature classifying telomerase and ALT immortalization reveals an hTERT regulatory network and suggests a mesenchymal stem cell origin for ALT. Oncogene. 28(43): 3765–74. doi: 10.1038/onc.2009.238.

Liu, C.C., D.L. Ma, T.D. Yan, X. Fan, Z. Poon, L.F. Poon et al. 2016. Distinct responses of stem cells to telomere uncapping-a potential strategy to improve the safety of cell therapy. Stem Cells. doi: 10.1002/stem.2431.

Meeker, A.K., J.L. Hicks, C.A. Iacobuzio-Donahue, E.A. Montgomery, W.H. Westra, T.Y. Chan et al. 2004. Telomere length abnormalities occur early in the initiation of epithelial carcinogenesis. Clin. Cancer Res. 10(10): 3317–26. doi: 10.1158/1078-0432.CCR-0984-03.

Mender, I., S. Gryaznov, Z.G. Dikmen, W.E. Wright and J.W. Shay. 2015. Induction of telomere dysfunction mediated by the telomerase substrate precursor 6-thio-2'-deoxyguanosine. Cancer Discov. 5(1): 82–95. doi: 10.1158/2159-8290.CD-14-0609.

Minamino, T., S.A. Mitsialis and S. Kourembanas. 2001. Hypoxia extends the life span of vascular smooth muscle cells through telomerase activation. Mol. Cell. Biol. 21(10): 3336–42. doi: 10.1128/MCB.21.10.3336-3342.2001.

Mitchell, J.R., E. Wood and K. Collins. 1999. A telomerase component is defective in the human disease dyskeratosis congenita. Nature. 402(6761): 551–5. doi: 10.1038/990141.

Nakamura, K., E. Furugori, Y. Esaki, T. Arai, M. Sawabe, I. Okayasu et al. 2000. Correlation of telomere lengths in normal and cancers tissue in the large bowel. Cancer Lett. 158(2): 179–84.

Nettleton, J.A., A. Diez-Roux, N.S. Jenny, A.L. Fitzpatrick and D.R. Jacobs, Jr. 2008. Dietary patterns, food groups, and telomere length in the Multi-Ethnic Study of Atherosclerosis (MESA). Am. J. Clin. Nutr. 88(5): 1405–12.

O'Sullivan, J.N., M.P. Bronner, T.A. Brentnall, J.C. Finley, W.T. Shen, S. Emerson et al. 2002. Chromosomal instability in ulcerative colitis is related to telomere shortening. Nat. Genet. 32(2): 280–4. doi: 10.1038/ng989.

O'Sullivan, J., R.A. Risques, M.T. Mandelson, L. Chen, T.A. Brentnall, M.P. Bronner et al. 2006. Telomere length in the colon declines with age: a relation to colorectal cancer? Cancer Epidemiol. Biomarkers Prev. 15(3): 573–7. doi: 10.1158/1055-9965.EPI-05-0542.

Oh, H., G.E. Taffet, K.A. Youker, M.L. Entman, P.A. Overbeek, L.H. Michael et al. 2001. Telomerase reverse transcriptase promotes cardiac muscle cell proliferation, hypertrophy, and survival. Proc. Natl. Acad. Sci. USA. 98(18): 10308–13. doi: 10.1073/pnas.191169098.

Oh, H., S.C. Wang, A. Prahash, M. Sano, C.S. Moravec, G.E. Taffe et al. 2003. Telomere attrition and Chk2 activation in human heart failure. Proc. Natl. Acad. Sci. USA. 100(9): 5378–83. doi: 10.1073/pnas.0836098100.

Opresko, P.L. and J.W. Shay. 2016. Telomere-associated aging disorders. Ageing Res. Rev. doi: 10.1016/j.arr.2016.05.009.

Ornish, D., J. Lin, J.M. Chan, E. Epel, C. Kemp, G. Weidner et al. 2013. Effect of comprehensive lifestyle changes on telomerase activity and telomere length in men with biopsy-proven low-risk prostate cancer: 5-year follow-up of a descriptive pilot study. Lancet Oncol. 14(11): 1112–20. doi: 10.1016/S1470-2045(13)70366-8.

Paul, L. 2011. Diet, nutrition and telomere length. J. Nutr. Biochem. 22(10): 895–901. doi: 10.1016/j.jnutbio.2010.12.001.

Rajaraman, S., J. Choi, P. Cheung, V. Beaudry, H. Moore and S.E. Artandi. 2007. Telomere uncapping in progenitor cells with critical telomere shortening is coupled to S-phase progression *in vivo*. Proc. Natl. Acad. Sci. USA. 104(45): 17747–52. doi: 10.1073/pnas.0706485104.

Ramunas, J., E. Yakubov, J.J. Brady, S.Y. Corbel, C. Holbrook, M. Brandt et al. 2015. Transient delivery of modified mRNA encoding TERT rapidly extends telomeres in human cells. FASEB J. 29(5): 1930–9. doi: 10.1096/fj.14-259531.

Reddel, R.R. 2003. Alternative lengthening of telomeres, telomerase, and cancer. Cancer Lett. 194(2): 155–62.

Risques, R.A., T.L. Vaughan, X. Li, R.D. Odze, P.L. Blount, K. Ayub et al. 2007. Leukocyte telomere length predicts cancer risk in Barrett's esophagus. Cancer Epidemiol. Biomarkers Prev. 16(12): 2649–55. doi: 10.1158/1055-9965.EPI-07-0624.

Risques, R.A., L.A. Lai, T.A. Brentnall, L. Li, Z. Feng, J. Gallaher et al. 2008. Ulcerative colitis is a disease of accelerated colon aging: evidence from telomere attrition and DNA damage. Gastroenterology. 135(2): 410–8. doi: 10.1053/j.gastro.2008.04.008.

Roberts, R.O., L.A. Boardman, R.H. Cha, V.S. Pankratz, R.A. Johnson, B.R. Druliner et al. 2014. Short and long telomeres increase risk of amnestic mild cognitive impairment. Mech. Ageing Dev. 141-142: 64–9. doi: 10.1016/j.mad.2014.10.002.

Rode, L., B.G. Nordestgaard and S.E. Bojesen. 2015. Peripheral blood leukocyte telomere length and mortality among 64,637 individuals from the general population. J. Natl. Cancer Inst. 107(6): djv074. doi: 10.1093/jnci/djv074.

Salvati, E., C. Leonetti, A. Rizzo, M. Scarsella, M. Mottolese, R. Galati et al. 2007. Telomere damage induced by the G-quadruplex ligand RHPS4 has an antitumor effect. J. Clin. Invest. 117(11): 3236–47. doi: 10.1172/JCI32461.

Samani, N.J., R. Boultby, R. Butler, J.R. Thompson and A.H. Goodall. 2001. Telomere shortening in atherosclerosis. Lancet. 358(9280): 472–3. doi: 10.1016/S0140-6736(01)05633-1.

Samper, E., J.M. Flores and M.A. Blasco. 2001. Restoration of telomerase activity rescues chromosomal instability and premature aging in Terc-/- mice with short telomeres. EMBO Rep. 2(9): 800–7. doi: 10.1093/embo-reports/kve174.

Scheinberg, P., J.N. Cooper, E.M. Sloand, C.O. Wu, R.T. Calado and N.S. Young. 2010. Association of telomere length of peripheral blood leukocytes with hematopoietic relapse, malignant transformation, and survival in severe aplastic anemia. JAMA. 304(12): 1358–64. doi: 10.1001/jama.2010.1376.

Shammas, M.A. 2011. Telomeres, lifestyle, cancer, and aging. Curr. Opin. Clin. Nutr. Metab. Care. 14(1): 28–34. doi: 10.1097/MCO.0b013e32834121b1.

Smogorzewska, A. and T. de Lange. 2004. Regulation of telomerase by telomeric proteins. Annu. Rev. Biochem. 73: 177–208. doi: 10.1146/annurev.biochem.73.071403.160049.

Song, Z., G. von Figura, Y. Liu, J.M. Kraus, C. Torrice, P. Dillon et al. 2010. Lifestyle impacts on the aging-associated expression of biomarkers of DNA damage and telomere dysfunction in human blood. Aging Cell. 9(4): 607–15. doi: 10.1111/j.1474-9726.2010.00583.x.

Tatsumoto, N., E. Hiyama, Y. Murakami, Y. Imamura, J.W. Shay, Y. Matsuura et al. 2000. High telomerase activity is an independent prognostic indicator of poor outcome in colorectal cancer. Clin. Cancer Res. 6(7): 2696–701.

Thomas, P., N.J. O' Callaghan and M. Fenech. 2008. Telomere length in white blood cells, buccal cells and brain tissue and its variation with ageing and Alzheimer's disease. Mech. Ageing Dev. 129(4): 183–90. doi: 10.1016/j.mad.2007.12.004.

Tiainen, A.M., S. Mannisto, P.A. Blomstedt, E. Moltchanova, M.M. Perala, N.E. Kaartinen et al. 2012. Leukocyte telomere length and its relation to food and nutrient intake in an elderly population. Eur. J. Clin. Nutr. 66(12): 1290–4. doi: 10.1038/ejcn.2012.143.

Townsley, D.M., B. Dumitriu, D. Liu, A. Biancotto, B. Weinstein, C. Chen et al. 2016. Danazol treatment for telomere diseases. N. Engl. J. Med. 374(20): 1922–31. doi: 10.1056/NEJMoa1515319.

Tsakiri, K.D., J.T. Cronkhite, P.J. Kuan, C. Xing, G. Raghu, J.C. Weissler et al. 2007. Adult-onset pulmonary fibrosis caused by mutations in telomerase. Proc. Natl. Acad. Sci. USA. 104(18): 7552–7. doi: 10.1073/pnas.0701009104.

Valdes, A.M., T. Andrew, J.P. Gardner, M. Kimura, E. Oelsner, L.F. Cherkas et al. 2005. Obesity, cigarette smoking, and telomere length in women. Lancet. 366(9486): 662–4. doi: 10.1016/S0140-6736(05)66630-5.

Valls, C., C. Pinol, J.M. Rene, J. Buenestado and J. Vinas. 2011. Telomere length is a prognostic factor for overall survival in colorectal cancer. Colorectal. Dis. 13(11): 1265–72. doi: 10.1111/j.1463-1318.2010.02433.x.

Vulliamy, T., A. Marrone, F. Goldman, A. Dearlove, M. Bessler, P.J. Mason et al. 2001. The RNA component of telomerase is mutated in autosomal dominant dyskeratosis congenita. Nature. 413(6854): 432–5. doi: 10.1038/35096585.

Vulliamy, T., A. Marrone, R. Szydlo, A. Walne, P.J. Mason and I. Dokal. 2004. Disease anticipation is associated with progressive telomere shortening in families with dyskeratosis congenita due to mutations in TERC. Nat. Genet. 36(5): 447–9. doi: 10.1038/ng1346.

Walne, A.J., T. Vulliamy, R. Beswick, M. Kirwan and I. Dokal. 2008. TINF2 mutations result in very short telomeres: analysis of a large cohort of patients with dyskeratosis congenita and related bone marrow failure syndromes. Blood. 112(9): 3594–600. doi: 10.1182/blood-2008-05-153445.

Wiemann, S.U., A. Satyanarayana, M. Tsahuridu, H.L. Tillmann, L. Zender, J. Klempnauer et al. 2002. Hepatocyte telomere shortening and senescence are general markers of human liver cirrhosis. FASEB J. 16(9): 935–42. doi: 10.1096/fj.01-0977com.

Wong, L.S., R.A. de Boer, N.J. Samani, D.J. van Veldhuisen and P. van der Harst. 2008. Telomere biology in heart failure. Eur. J. Heart Fail. 10(11): 1049–56. doi: 10.1016/j.ejheart.2008.08.007.

Wright, W.E. and J.W. Shay. 1992. The two-stage mechanism controlling cellular senescence and immortalization. Exp. Gerontol. 27(4): 383–9.

Wu, X., C.I. Amos, Y. Zhu, H. Zhao, B.H. Grossman, J.W. Shay et al. 2003. Telomere dysfunction: a potential cancer predisposition factor. J. Natl. Cancer Inst. 95(16): 1211–8.

Yamaguchi, H., G.M. Baerlocher, P.M. Lansdorp, S.J. Chanock, O. Nunez, E. Sloand et al. 2003. Mutations of the human telomerase RNA gene (TERC) in aplastic anemia and myelodysplastic syndrome. Blood. 102(3): 916–8. doi: 10.1182/blood-2003-01-0335.

Yamaguchi, H., R.T. Calado, H. Ly, S. Kajigaya, G.M. Baerlocher, S.J. Chanock et al. 2005. Mutations in TERT, the gene for telomerase reverse transcriptase, in aplastic anemia. N. Engl. J. Med. 352(14): 1413–24. doi: 10.1056/NEJMoa042980.

Yi, B.Q., B. Zhao and Z.J. Wang. 2011. Comparison of clinicopathological features and hRad21 expression between telomerase-dependent and telomerase-independent colorectal cancer. Hepatogastroenterology. 58(107-108): 785–9.

Index

For Product Safety Concerns and Information please contact our EU representative GPSR@taylorandfrancis.com Taylor & Francis Verlag GmbH, Kaufingerstraße 24, 80331 München, Germany